Marine Mammals of California

NEW AND REVISED EDITION

California Natural History Guides: 29

Marine Mammals of California

NEW AND REVISED EDITION

**Robert T. Orr
and
Roger C. Helm**

**Illustrations by
Jacqueline Schonewald**

Robert T. Orr

UNIVERSITY OF CALIFORNIA PRESS
Berkeley   Los Angeles   London

CALIFORNIA NATURAL HISTORY GUIDES
Arthur C. Smith, *General Editor*

Advisory Editorial Committee:
Raymond F. Dasmann
Mary Lee Jefferds
Don MacNeill
Robert Ornduff

University of California Press
Berkeley, California
University of California Press, Ltd.
London, England

Library of Congress Cataloging-in-Publication Data
Orr, Robert Thomas, 1908–
    Marine Mammals of California / Robert T. Orr and
Roger C. Helm.—Rev. ed.
        p.   cm.—(California natural history guides ; 29)
    Bibliography: p.
    Includes index.
    ISBN 0-520-06535-2
    ISBN 0-520-06515-8 (pbk. : alk. paper)
    1. Marine mammals—California.   I. Helm, Roger C.
II. Title   III. Series
QL713.2.O77 1989
599.5'09794—dc19   88-20739
                                                            CIP

Printed in the United States of America
1 2 3 4 5 6 7 8 9

Contents

Acknowledgments

Acknowledgments for this book have to do mostly with the illustrations. We would like to thank the following persons: Pieter Folkens for his expert counsel and patient monitoring of the cetacean portraits; Marc Webber and Drs. George E. Lindsay and Stephen F. Bailey for their constructive criticism of same; Marc Webber, Dr. Stephen F. Bailey, S. Jonathan Stern, Isidore Szczepaniak, Stanley Minasian, and Vincent Lee for the loan of photographic material. Two published works were used extensively as sources of visual reference and inspiration: *Whales, Dolphins, and Porpoises of the Eastern North Pacific and Adjacent Arctic Waters,* by Leatherwood, Reeves, Perrin, and Evans (1982); and *The World's Whales,* by Minasian, Balcomb, and Foster (1984). Other sources are listed in the bibliography.

We would also like to thank the individuals and institutions who have provided us with the photographs reproduced as black-and-white or color plates. All photographs not otherwise credited are by the senior author.

Special thanks to Dr. James Cunningham for his assistance in the creation of labels for the figures and to Gerry Stockfleth for the typing of the manuscript, as well as to Peigin Barrett, Executive Director of the California Marine Mammal Center, for the authorization to sketch and photograph some of the patients. The staff of the California Academy of Sciences library and Department of Exhibits also offered much valuable assistance.

Drs. Kenneth S. Norris and Mark O. Pierson kindly reviewed portions of an earlier draft of the manuscript.

Introduction

Fossil history shows us that primitive mammals were terrestrial, but that subsequently several groups have adapted to a marine or aquatic life. Those most specialized for life in water are the cetaceans (whales, porpoises, and dolphins), the pinnipeds (seals, sea lions, and the Walrus), and the sirenians (manatees, dugongs, and sea cows). A few species belonging to typically terrestrial mammalian groups have also become specialized for an aquatic existence—the Beaver, Muskrat, and Nutria among rodents; the Mink, river otters, and Sea Otter among carnivores; and hippos. Of these species, however, only the Sea Otter of the North Pacific and the Chungungo, a related species occurring along the west coast of South America, are truly marine.

The early history of the Pacific Coast of North America was greatly influenced by its marine mammal populations. The discovery of the Sea Otter and the Northern Fur Seal around the middle of the eighteenth century by the expedition under the command of Vitus Bering led to the exploration of western North America by the Russians from Alaska south to California. In their search for the fur of these animals the Russians moved steadily southward until they reached the central California coast, where they built Fort Ross in Sonoma County in 1812. There, with their Aleut assistants, they took large numbers of otter and seal pelts from our coastal waters in the early part of the nineteenth century. American fur hunters also slaughtered Sea Otters and seals, so that the populations of marine fur bearers were vastly reduced prior to the beginning of the twentieth century. Subsequent legislative protection has finally resulted in an increase in the numbers of most species.

Whaling activities were also carried on along the California coast by the start of the nineteenth century, but the first shore whaling station was not established until 1851. This was at Fields Landing near Eureka, Humboldt County, where it operated off and on for over a hundred years. The principal whaling station in central California in the early part of the twentieth century was at Moss Landing on the shore of Monterey Bay. A whaling station operated in San Francisco Bay until 1971, when the U.S. Department of Commerce ordered a ban on all commercial whaling by U.S. vessels.

Whaling has now been discontinued by most nations, but there are still a few that persist in this activity. It is hoped that continued world pressure against the exploitation of whales for commercial purposes will result in the complete cessation of all commercial whaling very shortly.

This book is for those who are interested in living marine mammals. It is intended not only to aid in the identification of species, but also to provide the reader with some background on their distribution and natural history. The first edition of *Marine Mammals of California* was published in 1972. Since then much new information on cetaceans, pinnipeds, and Sea Otters has accumulated. A number of species that were then considered rare in California waters are now known to be common. Much has been learned about their migration, reproduction, food habits, and general behavior, and the ranges of some species have changed markedly. The authors have attempted to summarize some of the material from these technical studies and present it in a concise but readable form. More information may be obtained by consulting the list of selected references at the end of the book. Terms with which readers may not be familiar are defined in the glossary.

1 • WHERE TO OBSERVE MARINE MAMMALS

California provides a most favorable coastline for observing a number of kinds of marine mammals. The Gray Whale's migration route, from Arctic waters to the west coast of Baja California and back, comes so close to shore that individuals may be seen from land at many places. Favorable localities for viewing include Point St. George, Trinidad Head, much of the Mendocino and Sonoma County coast, Point Reyes, the southern San Mateo Coast (especially Pigeon Point), the Monterey Peninsula, Point Pinos, Point Sur, and Point Loma (see Fig. 1). Tour boats also take whale watchers out for ocean viewing. Most of these are to be found at Sausalito, San Francisco, and Princeton (near Half Moon Bay) in the San Francisco Bay region, and at Soquel, Monterey, Ventura, and San Diego. On such trips other cetaceans are frequently observed. Humpback, Minke, and Killer Whales are occasionally encountered, and in recent years it is fairly common to see groups of Blue Whales. The latter are most likely to be seen in autumn off Monterey, where there are deep submarine canyons close to shore. Upwelling waters from these canyons provide the nutrients needed by *krill*, the small planktonic organisms that the whales feed upon. There are also submarine canyons off La Jolla that attract deepwater species such as Blue, Sei, and Fin Whales at certain times of the year.

Among the smaller cetaceans, species that are often seen from observation boats include the Common Dolphin, the Pacific White-sided Dolphin, Dall's Porpoise, and the Bottlenose

1

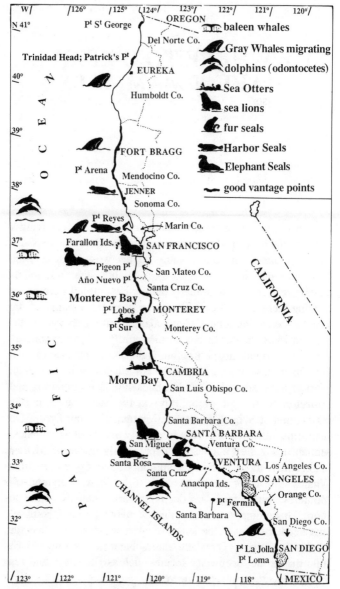

FIG. 1 The coast of California, showing major vantage points for observing marine mammals.

Dolphin. Most of these species seem to be attracted to small boats and will often ride the bow wave of the vessel.

Five species of pinnipeds commonly occur along the California coast, although their presence and abundance may vary with the season. One of these, the Northern Fur Seal, is a winter visitant (mostly females and immatures) as far south as San Diego. In 1968 a colony of about 100 Northern Fur Seals was discovered breeding on San Miguel Island. It is suspected that this colony started about four years previously. Today the population on this island during the breeding season has risen to over 10,000.

The California Sea Lion, the most abundant pinniped in California, breeds on the Channel Islands off southern California, but by the end of August the males have moved northward and may be seen or heard nearly everywhere along the coastline. Favorite viewing sites include Seal Rocks, just off the Cliff House in San Francisco, and the Monterey Breakwater. Steller's Sea Lion is much less common but breeds on Año Nuevo Island, San Mateo County, and on a few rocky offshore islands along the north coast of the state. This species formerly bred as far south as the Channel Islands, but only a few individuals are occasionally seen there now. In winter most of the Steller's Sea Lions have left the California coast and moved northward.

Harbor Seals were probably once common in suitable habitats along the entire California coast. Today, because of intense human utilization, they are essentially absent from southern California, except for the Channel Islands, but still fairly common from Point Conception north. They show preference for secluded bays, coves, rocky reefs, or sandy spits as well as the mouths of rivers such as the Russian River. They are relatively shy and tend to avoid areas where human beings are abundant.

Northern Elephant Seals, which were on the verge of extinction at the beginning of the twentieth century, are now common, seasonally, on certain offshore islands. The finest place for viewing these seals is at Año Nuevo Point in southern San Mateo County. A small colony that started breeding on the adjacent Año Nuevo Island in the winter of 1960–61 has so increased that it has spilled over onto the Point on the adjacent mainland. This area is presently under the administration of the California Department of Parks and Recreation. For a small fee, visitors are given guided tours of the area and the Elephant Seal colony.

Sea Otters presently occur from Santa Cruz south to San Luis Obispo County. They are most easily seen from points on the Monterey Peninsula including the Monterey wharf and Pacific Grove as well as at favorable spots on Seventeen-Mile Drive. Other excellent viewing sites are in Point Lobos State Park just south of the Carmel River.

A variety of live marine mammal species can also be viewed at California's excellent array of oceanariums and aquariums. Sea World in San Diego displays many species from Killer Whales and Bottlenose Dolphins to sea lions. The other major California oceanarium is Marine World/Africa USA, which is located in Vallejo, a few miles north of San Francisco. This park regularly displays Killer Whales, Bottlenose Dolphins, California Sea Lions, and Harbor Seals. Steinhart Aquarium in San Francisco maintains for public viewing Harbor Seals and dolphins.

In addition, several aquariums and museums, most notably the Monterey Bay Aquarium, the Whale Center in Oakland, the Gray Whale Museum at Point Loma, San Diego, and the San Diego Natural History Museum, have extensive displays devoted to marine mammals. The California Marine Mammal Center in Marin County, a volunteer organization, has been deeply involved in rehabilitating and releasing stranded young marine mammals. Visitors are welcome and tours of the facilities and the animals under their care are provided.

2 • WHALES, DOLPHINS, AND PORPOISES
Order Cetacea

Cetaceans are highly specialized aquatic mammals, believed to have developed from primitive terrestrial ancestors more than seventy million years ago. They have hairless, streamlined bodies with the forelimbs modified into flippers. External vestiges of hind limbs are lacking. The end of the tail is horizontally flattened into lateral extensions called *flukes*. A dorsal fin may or may not be present. The surface of the body lacks the usual *sebaceous glands* that lubricate the hair and skin of land mammals. Likewise sweat glands and scent glands are absent. Even the nipples of the mammary glands of the female are retracted into slits on either side of the genital opening when not in use. The ear lacks an external *pinna,* or flap, and the nostrils in all living cetaceans are situated on top of the head, rather posterior in position in all except the Sperm Whale. The nostril opening is called a *blowhole*. Externally there is a pair of blowholes in the baleen whales and a single opening in the toothed cetaceans. The upper and lower jaws are very elongate.

Cetaceans have developed many physiological adaptations to life in water. Beneath the skin is a thick layer of fat and connective tissue called *blubber,* which provides insulation against heat loss to the water. In some of the smaller species blubber may account for more than a third of the body weight. Fig. 2 provides a guide to the relative sizes of the cetaceans discussed in this chapter.

The respiratory and circulatory systems are specialized to withstand long periods under water. When a cetacean is submerged its pulse rate generally drops to about half of what it

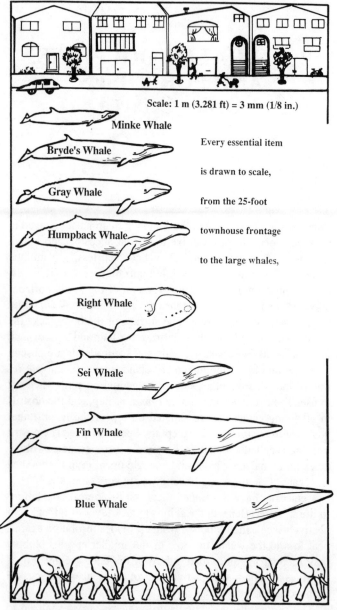

Scale: 1 m (3.281 ft) = 3 mm (1/8 in.)

Minke Whale

Bryde's Whale

Gray Whale

Humpback Whale

Right Whale

Sei Whale

Fin Whale

Blue Whale

Every essential item

is drawn to scale,

from the 25-foot

townhouse frontage

to the large whales,

FIG. 2 Relative sizes of cetaceans.

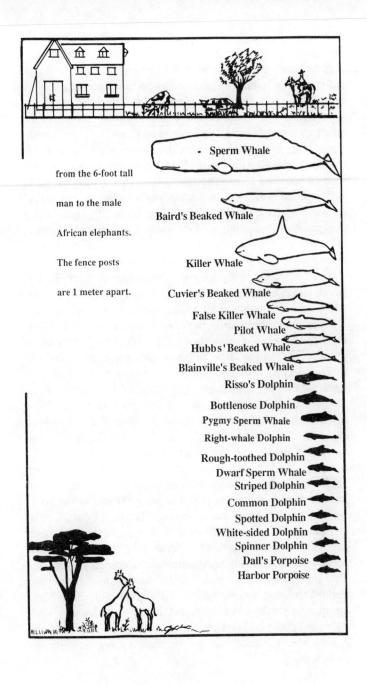

from the 6-foot tall

man to the male

African elephants.

The fence posts

are 1 meter apart.

Sperm Whale

Baird's Beaked Whale

Killer Whale

Cuvier's Beaked Whale

False Killer Whale

Pilot Whale

Hubbs' Beaked Whale

Blainville's Beaked Whale

Risso's Dolphin

Bottlenose Dolphin

Pygmy Sperm Whale

Right-whale Dolphin

Rough-toothed Dolphin

Dwarf Sperm Whale

Striped Dolphin

Common Dolphin

Spotted Dolphin

White-sided Dolphin

Spinner Dolphin

Dall's Porpoise

Harbor Porpoise

may be when the animal is at the surface. Circulatory shunts automatically reduce blood flow to certain parts of the body during dives, thereby ensuring a sufficient supply of oxygen to vital organs such as the brain and heart. Although many of the smaller cetaceans rarely go more than 15 to 20 seconds without coming to the surface to breathe, some of the larger deep-diving species like the Sperm Whale may stay under for more than an hour and dive to depths of over 915 m (3,000 ft).

Locomotion in cetaceans takes place primarily by vertical movement of the posterior part of the body. Strong muscular action accomplishes this, and the broad flukes provide the push that moves the body forward. The flippers function as stabilizers but provide essentially no propelling force.

Cetaceans apparently have no sense of smell. Sight is well developed in many species, although it is reduced in others, especially in river dolphins inhabiting turbid water.

In recent years it has been demonstrated that sounds covering a broad range of frequencies are produced by many kinds of cetaceans. In the large baleen whales these are thought to function largely for social communication. In the toothed cetaceans underwater sounds are also employed in orientation and food-finding. The sound pulses for this purpose are produced in nasal sacs and beamed in a specific direction. It has been suggested by some that the concave shape of the skull in the nasal area may serve as a sound reflector, and the so-called *melon* on the forehead focuses the beam. The echoes of such pulses are thought to be transmitted to the internal ears, at least in part, through the lower jaw. The posterior end of the lower jaw terminates next to the ear bones. The small external auditory canal also serves to direct sounds to the middle and inner ear. The tympanic bone, which houses the inner ear, is suspended from the skull by ligaments and housed in a cavity. This insulation permits sound to be directed to the inner ear through the auditory canal and ossicles of the middle ear. Since the specific gravity of the body and that of water are very similar, sound could enter the ear through any part of the body and disrupt accurate orientation if the auditory organs were not insulated.

The gestation period is long in all cetaceans, varying from 8 to 16 months depending on the species. The female usually has a single young, which grows rapidly on milk that has a 40 to 50 percent fat content.

It is customary to divide the living cetaceans into two major groups, the baleen or whalebone whales, or Mysticeti, and the toothed cetaceans, or Odontoceti. Some scientists place these groups in separate orders, but they are generally classified as suborders of the order Cetacea, which is the system followed here. Fig. 3 provides a comparison of the major anatomical features of the two groups.

BALEEN OR WHALEBONE WHALES
Suborder Mysticeti

The baleen whales are the giants of the sea. Their name is derived from the presence in the mouth of rows of long, flexible plates made of horny material much like fingernails. These plates, known as *baleen* or *whalebone,* hang down from each side of the upper jaw and may number several hundred per row. The inner margin of each plate is frayed into a hairlike fringe, and the plates combine to serve as food strainers. The food of these large whales consists principally of small, free-floating animals called *plankton,* which are especially abundant in polar waters as well as in temperate seas. Baleen whales also eat mollusks and small schooling fish, ranging from anchovies, sardines, and herring to saffron cod and capelin. A feeding whale takes in a great mouthful of water and then strains out the food organisms. Members of the family known as rorquals have accordion-pleated throats that are capable of great expansion to allow large quantities of water to be taken in at one time, thus increasing the efficiency of the straining mechanism.

The baleen plates are very long in the Bowhead and Right Whales, in the former measuring up to 4.5 m (15 ft) in length. In the Minke or Little Piked Whale, which is the smallest of the rorquals, the baleen measures about 30 cm (1 ft).

Baleen Whales

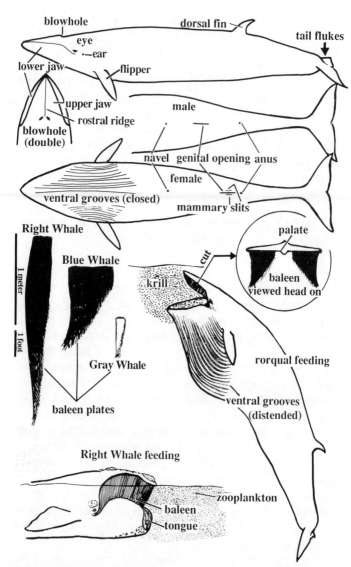

FIG. 3 Anatomical features of cetaceans. Above: baleen whales. Opposite page: toothed cetaceans.

Dolphin viewed from above

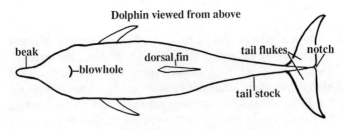

beak — blowhole — dorsal fin — tail flukes — notch — tail stock

Profile with outline of skeleton

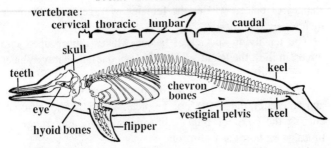

vertebrae: cervical — thoracic — lumbar — caudal

skull — teeth — eye — hyoid bones — flipper — chevron bones — vestigial pelvis — keel — keel

Detail of Risso's Dolphin flipper

Detail of human arm and hand

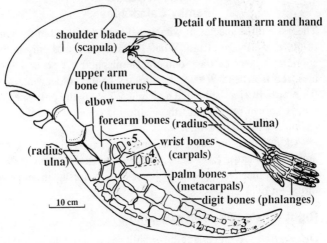

shoulder blade (scapula)
upper arm bone (humerus)
elbow
forearm bones (radius — ulna)
(radius ulna)
wrist bones (carpals)
palm bones (metacarpals)
digit bones (phalanges)

5
4
10 cm
1   2   3

Baleen Whales, unlike toothed whales, have two external blowholes on the top of the head. These cetaceans range in length from 6 m (20 ft) for the Pygmy Right Whale (*Caperea marginata*) of the Southern Hemisphere to 32 m (105 ft) for the Blue Whale (*Balaenoptera musculus*), which is found in all the oceans of the world. The record maximum weight for the latter species is estimated at about 145 metric tons. In all species of baleen whales the females average larger than the males.

The ten living species of baleen whales are grouped into three families: the Right Whales (Balaenidae), the Gray Whales (Eschrichtiidae), and the rorquals (Balaenopteridae). Seven of the ten species have been recorded along the California coast and an eighth, the Bryde's or Tropical Whale, is known to occur off the coast of Baja California and probably ranges at times as far north as California.

Right Whales
Family Balaenidae

There are three species of whales belonging to this family. One of these, the Pygmy Right Whale, is rare and found only in the subpolar waters of the Southern Hemisphere. A second species, the Bowhead Whale, is restricted to the colder waters of the Northern Hemisphere. The Right Whale is the only species that is known to occur off the California coast, where it is occasionally found in winter.

All members of the family are characterized by strongly arched jaws and a large head. The Pygmy Right Whale differs from the other two species in possessing a dorsal fin and two throat grooves.

Right Whale (*Balaena glacialis*)

Description. Average length for both sexes about 15 m (50 ft), with a maximum of 18 m (60 ft). Average adult weight 54 metric tons, with a maximum of 96 metric tons. Head large, constituting about 25 percent of body length. Body stocky, lacking any fin or ridge on the back. Horny growths or callosities present, principally in front of the blowhole, on the snout, and on the tip of the lower jaw. General color black, occasionally mottled with brown patches and irregular patches of white on the lower jaw and belly. Baleen plates, suspended from the arched upper jaw, over 2 m (7 ft) long and black in color (see Fig. 3).

FIG. 4 Right Whale (*Balaena glacialis*).

Distribution. Right Whales were once widely distributed in the temperate and subpolar waters of the world, but their numbers have been greatly reduced by whaling. Today they are confined to small local populations in parts of their former range. The species is separated into two distinct major populations that do not overlap, one in the Northern Hemisphere (*Balaena g. glacialis*) and one in the Southern Hemisphere (*B. g. australis*). There are a few more than 200 presently living in the North Pacific. They summer and feed in the Bering Sea and adjacent parts of the Arctic Ocean, and migrate southward in autumn along the eastern Pacific Ocean as far as the coast of northern Baja California. Recent sightings in California waters have been in the Santa Barbara Channel in 1981, Half Moon Bay in 1982, and La Jolla in February 1988.

Natural History. The name Right Whale is derived from the fact that the body is so rich in oil it floats even when dead; hence, this whale was considered by the early whalers to be the "right" whale to hunt. Also, it is relatively slow-moving and could easily be overtaken by the early wind-borne whaling ships. The total world population for this species is estimated to be about 3,500.

The food of the Right Whale consists largely of the small, shrimplike crustaceans called krill, which are strained out by the enormous fringed plates of baleen. The young are born in the winter, when the adults are in warmer water, and measure 5 to 6 m (16 to 20 ft) in length. The gestation period is 9 to 10 months. The single young suckles for about one year and often stays close to the mother for another two years.

Gray Whales
Family Eschrichtiidae

The family Eschrichtiidae contains a single species whose present range is limited to the eastern North Pacific Ocean. Another population formerly existed in the western Pacific, summering in the Sea of Okhotsk and wintering south of the Japanese island of Kyushu. It may now be extinct. Subfossil remains indicate that this species once occurred in the North Atlantic.

Gray Whale (*Eschrichtius robustus*)

Description. Average length for males 11 m (36 ft), maximum 15 m (50 ft); for females 12 m (39 ft), maximum 15 m (50 ft). Average weight for males 26 metric tons, for females 31 metric tons, with a maximum for the species of 34 metric tons. Head relatively small in contrast to the robust body, with a noticeable downward angle anterior to the blowhole when seen in profile. Dorsal fin absent. Posterior dorsal ridge on back, with a number of knobs or bumps anterior to the flukes. General body color dark gray with light mottling and extensive white streaking. Two to four throat grooves present. Baleen ivory in color and short, less than 40 cm (1.3 ft). Conspicuous patches of barnacles are visible on the back, head, and lower sides of the jaw. The lower jaw and the tip of the snout often possess numerous hairs. (See Pls. 1, 2, 4, 5, and 6.)

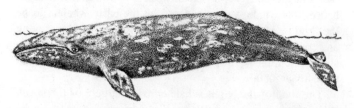

FIG. 5 Gray Whale (*Eschrichtius robustus*).

Distribution. Gray Whales summer in the Bering and Chukchi seas. In early autumn they begin moving south along the eastern Pacific coastline towards favored winter breeding areas in coastal Baja California. The southward migration begins in late September or early October. The whales typically travel in groups of two or more and generally stay within 10 km (6 mi) of the coastline. The migration is led by pregnant females, followed by nonpregnant females, males, and finally immatures. In most years the first individuals pass along the California coast in December, with peak numbers passing by in early January. During the northward movement, which begins in February, whales typically travel closer to shore in smaller, more widely spaced groups. Newly pregnant females are the first to head back to the northern feeding grounds, followed by adult males, nonbreeding females, and immatures. Females with young are the last to start north, beginning their migration in late March when the calves are ready to travel. The mother-young pairs travel at a leisurely speed, often very close to the shoreline, and pass along the California coast from April to June. This migration is the longest by any mammal. The round trip averages about 20,000 km (13,000 mi). It should be noted that in most years a few Gray Whales remain all summer long in coastal waters between California and British Columbia.

Natural History. Gray Whales were hunted intensively in the nineteenth century and were reduced in numbers almost to the point of extinction. They were easily slaughtered by early American whalers because of their slow movement and habit of wintering in shallow lagoons. With complete protection in the eastern Pacific since 1946, the species has made a remarkable comeback. Today the Gray Whale population is estimated to number somewhere between 15,000 and 20,000 whales, which is close to the estimated pre-exploitation level.

Gray Whales obtain most of their food during the summer months in the Arctic. It consists largely of amphipods and other crustaceans, along with some mollusks and other invertebrates. Many are secured on the bottom, where the whale uses its strong snout to stir up the sediment, from which the food is filtered by the baleen plates. In captivity, a very young female

Gray Whale named Gigi consumed 800 kg (1,800 lb) of squid a day and added weight at the remarkable rate of 1 kg (2 lb) per hour.

Mating, which occasionally occurs along the migration route, is mostly confined to Magdalena Bay and Scammon's, San Ignacio, and Black Warrior Lagoons along the west coast of Baja California. Usually a single 4.5-m (15-ft) 1.2-metric-ton calf is born after a 13-month gestation period. It is weaned the following summer in Arctic waters when 7 to 8 months of age and about 8.5 m (28 ft) long. Young whales often stay close to the protective mother for one or two additional years.

Gray Whales probably have good eyesight, and when near boats or the shoreline they occasionally rise vertically out of the water sufficiently high to scan the environment in all directions. This behavior, called *spyhopping,* may be maintained for as long as 30 seconds. Occasionally Gray Whales *breach,* a behavior more generally associated with Humpback Whales. Breaching for Gray Whales involves propelling the body about halfway out of the water then falling on the back with a great splash.

This species of whale is not considered to be particularly social. However, it is not uncommon to see groups of 4 to 5 whales traveling together, and groups of up to 20 have been recorded. In addition, as many a whaler belatedly discovered, Gray Whale mothers aggressively defend their calves. Killer Whales are the principal natural enemy of these whales, but they rarely attack adults.

Rorquals
Family Balaenopteridae

All Balaenopteridae are large to very large. The Blue Whale, considered the largest animal that has ever lived, has reached a length of 32 m (105 ft) and a weight of about 145 metric tons. Excluding the Humpback, all balaenopterids have proportionately small heads and trim, streamlined bodies adapted for speed. Due to the speed of these whales and the fact that they sink when killed, the early whalers were unable to capture them. However, modern whaling methods, which employ fast catcher boats continually supplying giant factory ships (essentially floating processing plants), have greatly reduced their numbers. All have a small dorsal fin and a throat and ventral thoracic region that is grooved or pleated to allow for great expansion when water is taken into the mouth to be strained for food.

Blue Whale (*Balaenoptera musculus*)

Description. Size very large, with a maximum recorded length of 32 m (105 ft). In the North Pacific males average 25 m (82 ft) and females 26 m (85 ft). Average weight ranges from 80 to 130 metric tons, with a record maximum weight of 145 metric tons. Body streamlined, with a small dorsal fin about 30 cm (1 ft) in height situated far back, and a sharp ridge extending from the blowholes to near the tip of the snout. Flippers small in proportion to body size. Numerous throat grooves that extend far back on the undersurface. Color blue-gray, somewhat mottled in temperate and warm waters, but yellowish from diatoms, especially on the ventral surface, in cold polar and subpolar waters. Baleen all black, with a maximum length of 1 m (39 in.).

FIG. 6 Blue Whale (*Balaenoptera musculus*).

Distribution. Originally worldwide, with summer feeding grounds in the cold subpolar waters where the small crustaceans known as krill occur in vast numbers. In the eastern North Pacific Blue Whales occur from around the Aleutian Islands in the Bering Sea south, seasonally, to tropical waters. Like other large whales, they migrate southward in autumn and north again in early summer. They usually travel in small groups of 2 to 20 individuals along the continental shelf–continental slope margin. They are seen most frequently off the California coast from July to January, occasionally occurring in Monterey Bay.

Natural History. The shy nature of Blue Whales, their great speed (they can travel at speeds in excess of 20 km/hr, or 11 knots), their general avoidance of shorelines, and the fact that they travel in small groups far from shore mean that population counts are difficult to obtain. The present estimates of world population range from 4,000 to 6,000, with from 1,000 to 1,500 of these occurring in the North Pacific. In former times this species was commonly called the Sulphur-bottom Whale because of the yellowish color acquired from diatoms attaching to its belly in Antarctic and Arctic waters during the summer feeding months.

From the commercial viewpoint, this was the most valuable of all whales because of its enormous size and high yield of oil. Consequently, intense harvesting efforts reduced the population from an estimate of over 200,000 to just a few thousand individuals in relatively few years. In the Antarctic alone, whaling vessels in 1930–31 took 30,000 individuals. No species of whale can withstand such high mortality rates for long.

Mating and calving in the North Pacific occur in the warmer waters during the winter months. Newborns measure between 6 and 7 m (19 and 23 ft) and weigh over 3 metric tons. The gestation period is from 11 to 12 months, and the calves nurse until about 7 months old. Females give birth to a young every two or three years.

Fin Whale (*Balaenoptera physalus*)

Description. Average length for North Pacific males 18 m (59 ft), for females 18.5 m (61 ft). Maximum length nearly

26 m (85 ft). Average weight for the species is from 35 to 45 metric tons, with a record maximum of 69.5 metric tons. Second in size only to the Blue Whale, which it resembles, the Fin Whale has a more wedge-shaped head and a larger dorsal fin, about 60 cm (2 ft) high. Numerous throat grooves extend back from the edge of the lower jaw to midbody. General body color is bluish gray or brownish above and light below. The head is asymmetrically pigmented, with the right side of the lower jaw, the baleen plates, the tongue, and occasionally the upper jaw being largely unpigmented and pallid, while the left side is very dark. Maximum length of baleen plates 90 cm (3 ft). In many specimens a light-colored chevron is apparent a few meters behind the blowhole when the whale is viewed from above.

FIG. 7 Fin Whale (*Balaenoptera physalus*).

Distribution. Fin Whales are widely distributed in the oceans of the world, but little is known about their movements. In the eastern North Pacific they are found at all seasons of the year from the Bering Sea to the Sea of Cortez in Mexico, with mating and calving occurring in the warmer southern waters. These are the most common of the rorquals off the California coast, although they are strictly pelagic. A small population is found during most of the year in the Sea of Cortez.

Natural History. The Fin Whale is not only the most widely distributed of the baleen whales, it is also, despite the heavy impact of whaling, one of the most common. The North Pacific population is now estimated at about 15,000, and the world population at 70,000 to 75,000.

Fin Whales can be gregarious, and pods consisting of a

dozen or more individuals are occasionally seen traveling together. Their food varies depending on availability, and includes small crustaceans, squid, and small schooling fish like sardines and anchovies. While feeding they will turn on their right side so that the pale coloration of that side is down, perhaps to the confusion of the prey. The very long, narrow body of the Fin Whale appears to be adapted for speed, as these whales have been reported to achieve 37 km/hr (20 knots) in short bursts.

The gestation and lactation periods are much the same as they are for the Blue Whale. Calves average about 6.5 m (21 ft) in length and weigh about 3.6 metric tons at birth.

Sei Whale (*Balaenoptera borealis*)

Description. Average length 13 m (43 ft), with females averaging slightly more than males. Maximum length 18.3 m (60 ft). Average weight 12 to 15 metric tons. Color and shape similar to the Blue Whale but much smaller, with the dorsal fin both actually and proportionally much taller (60 cm or 2 ft). Baleen black, lacking the asymmetry seen in the Fin Whale, and thinner than the baleen plates of any other rorqual. Throat grooves are numerous, but fewer than those of the larger whales, and they do not extend as far back on the belly.

FIG. 8 Sei Whale (*Balaenoptera borealis*).

Distribution. Worldwide in distribution. In the North Pacific this species has a range similar to that of the Fin Whale, but it tends to avoid the colder polar waters. There is no distinct migration pattern known. Peak numbers are found off the California coast from June to September.

Natural History. The proper pronunciation of *Sei,* a word of Norwegian derivation, is "Say." Population estimates for this species are no more than 20,000 in the North Pacific and less than 80,000 worldwide. This species has been actively sought by whalers since the decline of the Blue and Fin Whale populations.

Sei Whales tend to travel alone or in pairs, but may aggregate into larger pods when feeding. Like most members of this family, they feed on available zooplankton, squid, and small fish. Mating may occur anytime during the year, but is most frequent in late autumn and winter. Young are born a year later, and weaning occurs at about eight months. The young at birth measure 4.5 m (15 ft) and weigh about 900 kg (2,000 lb).

Bryde's or the Tropical Whale (*Balaenoptera edeni*) is very similar to the Sei Whale, but smaller and easily distinguished by the three ridges on top of the head (see Fig. 9). Bryde's Whale is also rarely recorded in the temperate waters where Sei Whales are most abundant.

FIG. 9 Bryde's Whale (*Balaenoptera edeni*).

Minke or Little Piked Whale (*Balaenoptera acutorostrata*)

Description. Average length 7 m (23 ft), with a maximum body length slightly over 10 m (33 ft). Average weight 6 to 8 metric tons. General color, like most rorquals, is blue-gray above and lighter below. A band of white across the middle of the upper side of the flippers is conspicuous. Other light patches frequently occur on the head or forepart of the body. Baleen very short and yellowish.

ments of this animal, named "Humphrey" by the media, were followed until it returned to the ocean on November 4. During these 25 days the animal moved up the bay and into the Sacramento River Delta, ending up in a narrow slough 116 km (70 mi) from the sea. Many attempts were made to direct Humphrey's course out of fresh water and back to the sea, but success was not achieved until a combination of underwater sounds was used. From a boat moving downstream in front of Humphrey, an amplified recording of Humpback Whales about to feed was broadcast. Simultaneously, from behind the whale, discordant sounds were produced by banging on pipes lowered into the water. This combination proved successful and Humphrey was last seen entering the ocean in the late afternoon of November 4. Humphrey was resighted August 16, 1986, in the Gulf of the Farallones, and again in the fall of 1987 and 1988, alive and well.

Natural History. The Humpback Whale is easily recognized by its stocky body, enormous flippers, and numerous knobby protuberances. Its habit of occasionally breaching or leaping partially or entirely out of the water and landing with a great splash on its back is also most distinctive.

Although Humpbacks have a relatively low oil yield, they have been hunted heavily over the years because of their size. Their numbers before human beings began exploiting them are estimated to have been around 100,000. The world population today is estimated at approximately 6,000, with at least 1,000 of these in the North Pacific. They are now completely protected.

Like other baleen whales, Humpback Whales obtain their food by straining crustaceans and schools of small fish, such as herring, captured in great gulps of water. Capture is accomplished in two ways by feeding whale groups. One is known as *bubble netting.* This involves a group of three or four individuals diving about 15 m (50 ft) below their prey, then slowly spiraling to the surface. As they spiral towards the surface they release a stream of air bubbles that effectively forms a net around the prey animals. This congregated mass of small fishes and other food items becomes trapped between the bubble net and the surface and is then consumed in great gulps of water. More commonly Humpbacks engage in *lunge feeding*. This involves moving very rapidly from below or from near the surface and lunging, with their mouths open, into schools of small fish.

Courtship, mating, and the birth of young all take place in warm water during the winter period. Sexual maturity is reached when less than 10 years of age. The gestation period for this species is from 10 to 11 months, and the newborn calves measure about 4.5 m (15 ft) and weigh about 1.3 metric tons. A calf will stay with its mother as long as a year but may be weaned as early as 6 months. Mating can take place shortly after the young are born, which allows some females to bear young annually.

One of the most interesting features of the Humpback Whale is its song. This is produced primarily, if not exclusively, by males, and it appears to have sexual significance. It is thought to be the longest continuous vocalization of any mammal. A single sequence may last for 15 minutes and be repeated without pause for hours at a time. Each local population has its own song, which often changes from year to year. These eerie underwater sounds can often be heard above the surface.

TOOTHED CETACEANS
Suborder Odontoceti

The toothed cetaceans or odontocetes include the sperm whales, the beaked whales, the Beluga and Narwhal, the Killer Whale, and the porpoises and dolphins. All have teeth in one or both jaws, the number varying from 2 to over 200. Although the great majority of toothed cetaceans are much smaller than the baleen whales, the food they consume is much larger and consists of fish and cephalopods.

The external nasal passages of odontocetes are joined to form a single blowhole, unlike the paired blowholes of the baleen whales. The single external orifice is usually crescentic or horizontal. In many species the rostral and frontal region of the head are greatly enlarged, with the Sperm Whale (*Physeter catodon*) being an extreme example. In others the mouth is extended forward, forming a long, narrow beak. This condition is a feature of most dolphins and beaked whales. The skulls of toothed whales are asymmetrical, in contrast to the symmetrical skulls of baleen whales.

The toothed cetaceans show great size variation. They range from the Sperm Whale, which may attain a length of 18 m (60 ft), to certain porpoises that measure only slightly more than 1.5 m (5 ft).

Currently there are about 74 living species of toothed cetaceans, grouped into at least six different families as follows: the sperm whales (Physeteridae), the Narwhal and Beluga (Monodontidae), the beaked whales (Ziphiidae), the river dolphins (Platanistidae), the ocean dolphins (Delphinidae), and the porpoises (Phocoenidae). About 18 kinds of odontocetes have been recorded along the California coast.

Sperm Whales
Family Physeteridae

This family contains two genera and three species of robust, stocky whales whose relatively large foreheads contain substantial quantities of a waxy oil called *spermaceti*. Spermaceti apparently plays a significant role in focusing the echolocation sounds produced by the nasal passages and perhaps the larynx. All members of this group are thought to be capable of deep dives in search of food. In the upper jaw teeth do not erupt past the gum line, but the underslung lower jaw contains from 7 to 29 pairs of well-developed teeth. The blowhole is situated somewhat to the left of the midline of the body, and the dorsal fin is either absent or, if present, it is *falcate,* which means hooked or curved like a sickle. Cephalopods (squid and octopus) constitute the principal food, but fish, crustaceans, and mollusks are also consumed.

Sperm Whale (*Physeter catodon*)

Description. Sperm Whales are the largest toothed whales. Males over 18 m (60 ft) long have been recorded, with most adult males measuring at least 15 m (50 ft). The smaller female averages 11 m (36 ft). Mature males typically weigh

40 metric tons, with females averaging less than 15 metric tons. The head of a Sperm Whale is blunt and massive, constituting one-third or more of the body length. There is no true dorsal fin, but a low-slung humped region is apparent in the latter one-third of the back, followed by numerous knucklelike bumps. The blowhole of Sperm Whales, unlike that of other cetaceans, is situated on the left upper tip of the head, resulting in a blow (exhalation) that extends forward at about a 45-degree angle. The flukes are distinctly triangular with a median notch. The predominant body coloration is dark brownish gray, with many individuals also showing a scattering of light-colored splotches. The area behind the head is heavily wrinkled, which gives this huge animal a shriveled appearance. The narrow underslung lower jaw is usually outlined in white and contains 18 to 29 very large conical teeth. The white-outlined upper jaw, in which teeth do not erupt past the gum line, contains deep sockets to accommodate the teeth of the lower jaw.

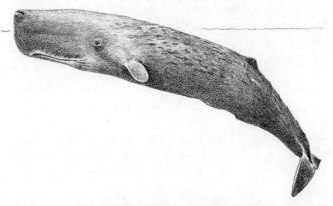

FIG. 12 Sperm Whale (*Physeter catodon*).

Distribution. All oceans. Sighted off California in every month except July and considered relatively common over continental slope waters from November to April. In general, Sperm Whales prefer deepwater areas where strong currents flow in opposite directions, such as can be found along the edges of oceanic trenches. (These areas, known as *shear zones,* tend to concentrate nutrients, are highly productive, and are therefore favorite feeding areas for a number of marine vertebrates.) The movement patterns and distribution of this whale are segregated based on age and sex. All animals tend to move north in spring and south in autumn, but males, particularly solitary bulls, typically move farther north and travel faster than females. Sperm Whales tend to occur in three distinct groupings: adult females with young and juvenile whales, occasionally accompanied by a mature male; groupings of young to mature bachelor males; and solitary mature bachelor males.

Natural History. Sperm Whales are gregarious, often traveling in pods of 30 or more, and they are commonly seen in association with other marine mammals such as Risso's Dolphins, Northern Right-whale Dolphins, Pacific White-sided Dolphins, Dall's Porpoises, and Northern Fur Seals. The tight cohesiveness of Sperm Whale groups probably contributes to their tendency to mass strand on occasion. Perhaps the most remarkable attribute of this species is its ability to dive rapidly to tremendous depths and stay submerged for over one hour. Based on the round suckerlike scars found on the head and body of Sperm Whales and analyses of stomach contents, it appears that these whales dive to great depths in order to capture giant squid and other deep-dwelling species.

Mating occurs primarily in spring and summer, with bulls possibly engaging in fierce battles to gain control of pods of adult females. Gestation takes 14 to 16 months, and weaning occurs at one and a half to two years of age. Thus, the calving interval is at least three years and maybe as long as five years. Newborn calves weigh 1 metric ton and measure about 4 m (13 ft).

This species has been a mainstay of the whaling industry since the 1700s because it contains large amounts of valuable spermaceti oil, which can be used in the manufacture of candles and some medicinal products. In modern times the take of Sperm Whales peaked in 1963, when over 30,000 whales were killed. Since that time the catch has fallen continuously. Present population estimates vary so widely they are functionally useless.

Pygmy Sperm Whale (*Kogia breviceps*)

Description. Length to 3.4 m (11 ft) and weight to 410 kg (900 lb). Most adults do not exceed 3 m (10 ft) and 360 kg (800 lb). The blunt head constitutes approximately 15 percent of body length. Jaws short, triangular, and placed behind the snout tip. Compared to most odontocetes, the blowhole is displaced slightly anterior and to the left. Flippers are short and broad. The small dorsal fin is falcate and placed behind the midpoint of the back. Back and flanks are steely gray, and the belly is a lighter gray with a tinge of pink. A gray-white bracket-shaped mark called a false gill extends between the eye and the flippers. The underslung jaw in combination with the false gill gives this species the appearance of a bottom-feeding shark. Twelve to 16 pairs of sharp recurved teeth line the lower jaw.

FIG. 13 Pygmy Sperm Whale (*Kogia breviceps*).

Distribution. Cosmopolitan in temperate, subtropical, and tropical waters.

Natural History. Pygmy Sperm Whales are rarely identified at sea, where they are typically seen hanging motionless at the surface with their tails pointing down. When disturbed, most seem to produce an orange-rust-colored defecation before sounding. Nearly all sightings involve five or fewer individuals. Primarily a bottom feeder, this species takes squid, fish, and crabs, possibly from considerable depths.

Most births occur between autumn and spring following 9 to 11 months of development. The 1.2-m (4-ft) newborn weighs 55 kg (120 lb) and probably nurses for at least 12 months.

Dwarf Sperm Whale (*Kogia simus*)

Description. The majority of adults measure 2.4 m (8 ft) and weigh around 150 kg (330 lb). Very large individuals can reach 2.75 m (9 ft) and 270 kg (600 lb). Body shape and form are similar to the Pygmy Sperm Whale, and it is only within this century that the Dwarf Sperm Whale has been considered a separate species. The characteristics peculiar to the Dwarf Sperm Whale are the midbody placement of its taller dorsal fin, the presence of a number of short throat grooves, fewer teeth in the lower jaw (7 to 12), and the occasional occurrence of vestigial teeth in the upper jaw.

FIG. 14 Dwarf Sperm Whale (*Kogia simus*).

Distribution. Unknown, but thought to be cosmopolitan from temperate to tropical waters.

Natural History. Group size typically seven or fewer. Consumes squid, crustaceans, and fish. Known to stay submerged for extended periods, and capable of diving to depths of more than 250 m (820 ft).

Reproductive biology is not known.

Beaked Whales
Family Ziphiidae

This group of small to medium-sized whales is represented by three genera and seven species in the eastern North Pacific. For the most part these whales are poorly known, as they seem to prefer deep oceanic waters (usually in excess of 1,000 m, or 3,300 ft) and typically submerge when approached. Nearly all members of this family have a well-defined beak that contains only one or two pairs of teeth in the lower jaw. The dorsal fin is triangular to falcate and set well back on the body. The throat contains a pair of grooves that often joins just below the tip of the lower jaw. All members are thought to be deep divers and harassed individuals have been known to *sound,* or dive deeply, for over one hour. Cephalopods constitute the bulk of their diet.

Most beaked whales occur predominantly in *continental slope waters,* which are waters between 200 and 2,000 m (660 and 6,600 ft) deep that lie between the point where the continental shelf ends and the deep ocean floor. However, Baird's Beaked Whale occurs regularly in *continental shelf waters,* which are less than 200 m (660 ft) deep.

Baird's Beaked Whale (*Berardius bairdii*)

Description. Baird's Beaked Whale is the largest member of the family Ziphiidae. Adult females average a bit over 11 m (37 ft), but specimens nearly 13 m (40 ft) long have been recorded. The adult male averages around 10 m (34 ft). Both sexes weigh in excess of 8.5 metric tons. This species is the only North Pacific beaked whale in which adult individuals may have two pairs of teeth visible above the gum line, and it is also the only species in which the teeth erupt in both sexes. The dorsal surface and flanks of the long narrow body varies from a dark bluish gray to brown, often covered with blotches or parallel whitish scratches. The ventral side is usually paler. The well-pronounced beak extends prominently from the bulbous forehead, and the dorsal fin is triangular in shape.

FIG. 15 Baird's Beaked Whale *(Berardius bairdii).*

Distribution. Found only in the North Pacific. Along the eastern Pacific seaboard these whales range from Alaska to Baja California. Seasonal north-south shifts in distribution probably occur, but the timing and extent of movement is poorly known. Whereas most beaked whales occur predominantly over the continental slope, this species is regularly found over continental shelf waters.

Natural History. The exact population status of Baird's Beaked Whale remains to be determined, but we do know it is the most abundant ziphiid off California. It is also fairly common in the western North Pacific, as shore-based fisheries off northern Japan have taken a few hundred every year since at least the 1940s. This species commonly travels in groups of 3 to 10 individuals, and sightings of up to 30 whales traveling together have been recorded. Deepwater fishes make up the bulk of the diet. The nature of certain scratches and scars found around the head and mouth suggests these whales might feed directly on the ocean floor.

Reproduction is slow, with one 4- to 5-m (15-ft) calf being born about every three years. Most calves are born in spring after a long gestation period.

Cuvier's Beaked Whale (*Ziphius cavirostris*)

Description. The larger female Cuvier's Beaked or Goosebeaked Whale, as it is also known, can exceed 7 m (23 ft) in length and weigh nearly 4.5 metric tons. The smaller male averages under 6 m (20 ft) and weighs around 3 metric tons. The dorsal fin is typically falcate and positioned on the latter one-third of the body. Body color is variable, ranging from reddish brown to umber. Numerous whitish blotches and parallel scratches give the body a mottled appearance. In both sexes the head is white, with the white extending farther back on the nape in older males. The beak is relatively short for this family and the mouth, when viewed from the side, turns distinctively upward at the rear, giving the head a goosebeaklike profile. The tip of the lower jaw contains a pair of conical teeth that erupt past the gum line in adult males only.

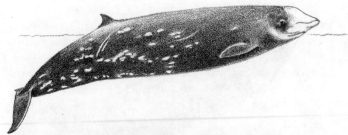

FIG. 16 Cuvier's Beaked Whale (*Ziphius cavirostris*).

Distribution. In all oceans except extreme northern and southern latitudes. In the eastern North Pacific its range extends from the southern Bering Sea to the equator. Little is known of the migration patterns, but these whales seem to move north and south in synchrony with the changing seasons. Off California they are uncommon in water shallower than 1,000 m (3,300 ft).

Natural History. Cuvier's Beaked Whale is considered by some experts to be one of the most abundant beaked whales in the world, but they are still seldom identified off California. The diet, as with most ziphiids, consists primarily of squid and deepwater fishes. This species commonly travels in groups of 2 to 15 individuals, and its behavioral repertoire includes an occasional breach. The scant amount of data collected suggests that newborn calves average about 2 to 3 m (6 to 10 ft) and that calving occurs year-round.

Genus *Mesoplodon*

Description. Considerable controversy surrounds the exact taxonomy of this rare group of small whales, as nearly all of our knowledge is based on a limited number of stranded individuals. At present, five members of this genus are recognized as occurring in the North Pacific: Hubbs' Beaked Whale (*Mesoplodon carlhubbsi*), Stejneger's Beaked Whale (*M. stejnegeri*), Blainville's Beaked whale (*M. densirostris*), Ginkgo-toothed Beaked Whale (*M. ginkgodens*), and Hector's Beaked Whale (*M. hectori*).

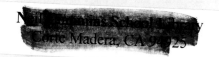

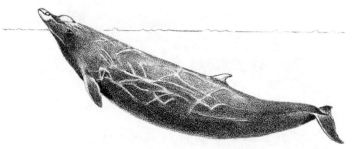

FIG. 17 Hubbs' Beaked Whale (*Mesoplodon carlhubbsi*), male.

Adults of all these species measure about 5 m (16.5 ft) and weigh around 1,500 kg (3,300 lb). The low falcate dorsal fin is always positioned in the latter one-third of the body, and the generally spindle-shaped body shows little discernible variation. Descriptions of color patterns have been as variable within species as among them. Many individuals are extensively scarred with white scratch lines and oval markings along the back and sides of the body. Both males and females of the genus *Mesoplodon* have a single pair of teeth, which only erupt through the gum line in the male.

FIG. 18 Blainville's Beaked Whale (*Mesoplodon densirostris*), female.

At present only adult males can be reliably identified by persons other than trained experts (see Figs. 17, 18, and 19 and Pl. 10). The adult male Hubbs' Beaked Whale has a raised white crown to the forehead in front of the blowhole, a white-tipped beak, and the apex of each tooth is behind its leading edge. In Stejneger's Beaked Whale neither the beak nor the crown area are white, and the apex of each tooth is in line with

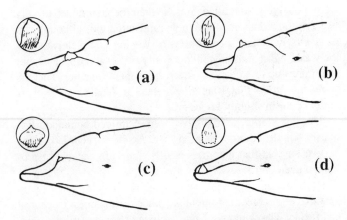

FIG. 19 Head detail of four male mesoplondont beaked whales recorded in California waters showing the development of the jaw and shape of the paired teeth. (a) Hubbs' Beaked Whale. (b) Blainville's Beaked Whale. (c) Gingko-toothed Beaked Whale. (d) Hector's Beaked Whale.

its leading edge. The lower jaw in Blainville's Beaked Whale is massive and obviously arched, and the single laterally flattened pair of teeth are angled forward. In addition, the forehead and beak of this species are depressed and nearly concave, which contrasts sharply with the bulging convex nature of this region in most cetaceans. As its common name implies, the single tooth that erupts in male Ginkgo-toothed Whales resembles the leaf of a Ginkgo tree. The outline of this tooth might also be described as similar to that of a human breast. Until this decade, when a couple of stranded specimens were recovered in southern California, Hector's Beaked Whale was not known to occur in the North Pacific. By *Mesoplodon* standards this species has a short beak, but its most distinctive characteristic is that the teeth in males are situated near the tip of the lower jaw. In all other *Mesoplodon* species in the North Pacific the teeth are set several inches back from the jaw tip.

Distribution and Natural History. Little is known regarding the distribution and natural history of these species. It seems likely that they primarily occur well offshore and generally tend to avoid vessels. On the few occasions when live animals were positively identified at sea, their behavior has usually been described as inconspicuous and lethargic, although, on occasion, they have been observed leaping completely out of the water. Most sightings involve only two or three whales, but sometimes up to a dozen or more are sighted together. There is some indication that Blainville's Beaked Whale may consistently occur in slightly larger groups.

The few stomachs that have been examined contained primarily squid, although salmon has also been recorded. Most experts suspect that members of this genus are capable of prolonged deep dives.

Recent strandings of newborn Hubbs' Beaked Whales suggest that some breeding may take place off the coast of central California. The weights of two stranded neonates were as follows: a male, 131 kg (290 lb); a female, 156 kg (344 lb).

Ocean Dolphins
Family Delphinidae

This is by far the largest family of cetaceans, and it contains most of the animals typically thought of as dolphins. A distinct beak is often present and, excluding the Northern Right-whale Dolphin, all members possess a prominent dorsal fin. In all species except Risso's Dolphin, teeth are present in both upper and lower jaws, and they tend to be conical, sharp, and well-developed. General body form and life-style are variable.

Killer Whale (*Orcinus orca*)

Description. Killer Whales, also called Orcas, are the largest members of the family Delphinidae. Adult males measure over 9 m (30 ft) and weigh in the vicinity of 8 metric tons. Adult females average 7 m (23 ft) and 4 metric tons. The distinctive black and white markings of this whale are known to visitors of oceanariums the world over. The dorsal surface and flanks are glossy black with a light gray saddle just behind the dorsal fin. The throat, belly, and flank area behind and below the dorsal fin are glistening white, as is a small oval patch that extends posteriorly from above the eye. The erect triangular dorsal fin of adult males, which can measure up to 1.8 m (6 ft), contrasts sharply with the 60-cm (2-ft) moderately falcate fin of females. Each jaw usually contains 11 strong sharp conical teeth that are ideally suited for grasping and tearing prey. (See Pl. 3.)

FIG. 20 Killer Whale (*Orcinus orca*).

Distribution. All oceans but most abundant in cooler temperate areas. In the eastern North Pacific they are most common from Monterey Bay north to the Aleutian Islands. They occur near shore, but groups are mostly sighted in both deep ocean and inlet areas. In some regions, most notably Puget Sound in the state of Washington, these whales are year-round residents. Off California, Orcas are known to shift their distribution seasonally in response to changing food resources.

Natural History. Many Californians are first introduced to Orcas while visiting a local oceanarium. These highly intelligent and trainable cetaceans survive well in captivity, and their displays of power, agility, and sheer size are fascinating. Orcas are the top predators in the ocean, consuming nearly anything that crosses their path.

In the wild this highly social species is commonly found in pods of 5 to 30 animals. Most pods consist of related individuals, and some of these family units stay together for decades. Orcas are capable of producing a large variety of sounds, and pods from different areas have different dialects. The large repertoire of sounds apparently plays an important role in maintaining group cohesion when a pod is spread out over several kilometers during a group hunt.

Calving peaks in fall and winter, and the interbirth interval is two years or more. Gestation lasts at least 12 months, and the period of maternal care appears to be protracted. At birth a calf measures 2.5 m (8 ft) and weighs about 180 kg (400 lb).

False Killer Whale (*Pseudorca crassidens*)

Description. Males average 5.5 m (18 ft) and weigh about 1.4 metric tons. Females average 4.7 m (15 ft) and 1.1 metric tons. Maximum size is nearly 6 m (20 ft) and 2 metric tons. The body is long and slender, the head is small and somewhat rounded. The mouth is large and the tip of the longer upper jaw has a swollen appearance. The slender, falcate dorsal fin is centrally placed, and the flippers have a distinct hump on their leading edge. Besides light scarring, the overall blackish body coloring is interrupted only on the neck and belly by a light gray blaze. Each jaw contains 8 to 11 stout teeth. (See Pl. 11.)

FIG. 21 False Killer Whale (*Pseudorca crassidens*).

Distribution. A cosmopolitan offshore species that avoids polar seas. Unusual north of southern California. Migration patterns are not well known, but some populations apparently move north in spring and summer.

Natural History. Typically found in groups of more than 100, and often seen traveling with other odontocetes. The cohesiveness of certain groups is demonstrated by their tendency to mass strand occasionally. Captive False Killer Whales have proved to be fast learners, agile, and accomplished aerial leapers. They are also known to be emotional, volatile, and aggressive at times. Their diet primarily consists of squid and large fish such as various species of tuna, Pacific Bonito, and Dolphinfish (Mahi-mahi). Pelagic dolphins may also be consumed on occasion.

Reproductive biology is poorly known. Calves are seen throughout the year, and newborns weigh around 80 kg (175 lb) and measure about 1.5 m (5 ft).

Short-finned Pilot Whale (*Globicephala macrorhynchus*)

Description. Mature males average 5 m (16 ft) with a maximum of 5.9 m (19 ft), and females average around 3.5 m (11 ft). Maximum weight for males is 3 metric tons, which is nearly double the maximum weight recorded for females. General body coloration is dark brown to black with a gray-pigmented saddle posterior to the dorsal fin. With its broad base and distinctly curved tip, the low-slung dorsal fin is an excellent field mark for identifying these whales at a distance. Other distinctive characteristics of this whale are its bulbous

FIG. 22 Short-finned Pilot Whale (*Globicephala macrorhynchus*).

bulging forehead and its sickle-shaped pectoral fins. The ventral surface bears a whitish patch of variable length that extends posteriorly from the throat along the midline of the body. Each jaw contains 7 to 9 stout conical teeth. (See Pls. 7 and 11.)

Distribution. Most common from Point Conception south to Central America and across the Pacific to the Hawaii Archipelago. Rare north of Monterey Bay. Throughout their range, Short-finned Pilot Whales are found both nearshore and far offshore. The patchiness of their distribution suggests that these whales are divided into several distinct stocks or populations. Migration patterns are not well researched, but we do know that off southern California some whales move closer to shore in late winter or early spring to feed on squid that have concentrated near shore to spawn.

Natural History. Short-finned Pilot Whales off California tend to travel in groups of 20 to 40 individuals. As with many of the other social cetaceans, these whales commonly associate with other species like the Bottlenose, Common, and Northern Right-whale Dolphins. They emit a variety of vocalizations, and in captivity they have proven to be quick learners capable of a wide variety of behaviors. Both wild and captive whales have demonstrated interindividual helping behavior such as attending to and supporting injured or even deceased companions. This species is also well known for its voracious consumption of squid and propensity to strand en masse.

Reproductive patterns of Short-finned Pilot Whales in the North Pacific are poorly known. In other areas these whales produce a single 1.4-m (5-ft) calf about every three years. This slow reproductive rate is a consequence of a gestation period that lasts 15 to 16 months and a nursing period that lasts for well over a year. Calving probably occurs throughout the year.

Rough-toothed Dolphin (*Steno bredanensis*)

Description. The slightly larger male averages 2.4 m (8 ft) and 130 kg (285 lb). Maximum reported size is 2.75 m (9 ft) and 160 kg (350 lb). The narrow, light-colored beak is elongated and tapers smoothly from the forehead. Dorsal fin moderately tall, slightly falcate, and placed midbody. Flippers long.

Dorsal surface and upper part of flanks bluish gray to purplish black. Ventral surface and lower flanks pinkish white. Yellowish white spots and streaks scattered over body. Twenty to 24 pairs of well-developed teeth in each jaw. Tooth crown is rough, as the surface is striated with fine vertical ridges. (See Pl. 8.)

FIG. 23 Rough-toothed Dolphin (*Steno bredanensis*).

Distribution. Tropical and subtropical waters. Although a few specimens have stranded on California beaches, this species is considered rare in the temperate waters off this coast.

Natural History. Infrequently identified at sea, and its habits are not well known. Gregarious, associates with other species, and is often sighted in groups of 50 or more. Will ride the bow wave of boats, and burst speeds are reported to exceed 28 km/hr (15 knots). In captivity this species is highly inventive. Diet consists of fish and cephalopods.

Reproductive biology is not known.

Pacific White-sided Dolphin (*Lagenorhynchus obliquidens*)

Description. Most adult Pacific White-sided Dolphins measure about 2 m (7 ft) and weigh approximately 90 kg (200 lb). Very large individuals can exceed 2.3 m (7.5 ft) and weigh nearly 150 kg (330 lb). When viewed from above, the dark blue-gray to black back of these dolphins contrasts strongly with gray-white stripes that extend posteriorly along each side of the body from the blowhole to just in front of the flukes. The flanks are light in the front and rear and dark in the middle. They are distinctly separated from the white belly by a black line running the entire length of the body. The head is conical with a short but distinct beak. The lightly colored rear two-thirds of the strongly falcate dorsal fin is an excellent field mark for identifying these dolphins from a boat. Approximately 25 sharp, slightly inwardly curved teeth line each jaw. (See Pl. 13.)

FIG. 24 Pacific White-sided Dolphin (*Lagenorhynchus obliquidens*).

Distribution. Restricted to the temperate North Pacific, these dolphins range from Japan to the Gulf of Alaska to Baja California and are one of the most abundant cetaceans in this area. Seasonal inshore-offshore and north-south movements occur primarily in response to changing food availability. It also seems likely that the inshore movements of late summer and early autumn coincide with peaks in reproductive activity. This species tends to reach its greatest abundance along the California coast in September and October.

Natural History. Pacific White-sided Dolphins are extremely gregarious, often occurring in schools of 100 or more, and groups in excess of 1,000 are fairly common. This species often associates with other marine mammals, particularly Risso's and Northern Right-whale Dolphins; Blue, Humpback, and Gray Whales; and some pinniped species.

Pacific White-sided Dolphins are active and rapid swimmers, and they commonly ride the bow wave of boats. They are frequently seen chasing one another while making high-speed dives, zigs, zags, and occasional end-over-end aerial flips. One of the most exciting behavioral patterns of this species occurs when a hundred or more individuals repeatedly leap several feet into the air and slam resoundingly back into the ocean. As with many offshore marine mammal species, the diet of Pacific White-sided Dolphins consists of schooling fish and squid captured at night.

Mating peaks in autumn, with the young being born the following summer after a gestation period of 9 to 10 months. Newborn calves measure 1 m (3 ft) and weigh 15 kg (33 lb). Mothers with unweaned young are sighted most frequently in groups that exceed 50 individuals.

Common Dolphin (*Delphinus delphis*)

Description. Most Common Dolphins measure about 2.2 m (7 ft) and weigh around 80 kg (180 lb). Very large individuals can measure 2.6 m (8.5 ft) and weigh as much as 135 kg (300 lb). On the average males are larger than females, but the difference is not great. This sleek, spindle-shaped dolphin is distinctly pigmented. The uniformly dark dorsal and light ventral region contrasts sharply with the elaborately crisscrossed patterns on the flanks. Anteriorly the flanks are yellowish tan

FIG. 25 Common Dolphin (*Delphinus delphis*).

above and white below, with the posterior portion being gray above and darker below. Bold dark stripes extend forward from the flippers and from a black patch surrounding each eye. One or two light gray stripes run posteriorly from the lower jaw. The dark jaws extend prominently past the forehead, forming a distinctive beak. Between 40 and 57 sharp conical teeth line each jaw. The dorsal fin is tall and the shape varies from triangular to falcate. (See Pls. 9 and 12.)

Distribution. Tropical and warm temperate oceanic waters. In the eastern North Pacific they typically range from the equator to southern California, showing a distinct preference for water temperatures between $10°$ and $28°$ C ($50°$ and $82°$ F). In extreme warm-water years these dolphins will move as far north as British Columbia. Common Dolphins appear to be divided into many subpopulations, some of which occur nearshore and others that occur primarily well offshore. In and around the waters off southern California, where this species is the most abundant cetacean, groups of dolphins exhibit seasonal shifts in their distribution as they follow food resources along submarine ridges. Although this species is present off southern California year-round, peak numbers occur in January, June, September, and October.

Natural History. A very active and social species, the Common Dolphin is frequently sighted in groups of several hundred individuals leaping high into the air as they travel. These dolphins readily ride the pressure wave in front of a whale or a boat, sometimes for a considerable period of time.

Feeding probably occurs primarily in the late evening and at night when deepwater fish and squid rise towards surface waters. When diving, Common Dolphins frequently stay submerged for 5 to 8 minutes and may reach depths in excess of 250 m (820 ft). They have been clocked at speeds greater than 30 km/hr (16 knots).

Calving peaks in spring and autumn. Gestation lasts 10 to 11 months and lactation 5 to 6 months. Although this species is apparently capable of calving in successive years, it appears more commonly to calve every other year. At birth, a calf measures about 80 cm (2.6 ft).

Bottlenose Dolphin (*Tursiops truncatus*)

Description. Bottlenose Dolphins are probably the most familiar cetacean in the world as a result of their numerous television and movie appearances and their widespread use as a performing sea creature. This robust dolphin averages around 3 m (10 ft) and 250 kg (550 lb), with particularly large individuals reaching nearly 4 m (13 ft) and 275 kg (600 lb). Body coloration is generally medium gray above with pale gray flanks and a lighter, sometimes pinkish belly. In certain areas the belly region of older animals may be spotty. The jaws extend forward of the forehead to form a medium-length beak. Although the lower jaw is slightly longer than the upper, it usually contains fewer teeth (18 to 24, vs. 20 to 26 in the upper jaw). The strong arching of the back in combination with the midbody placement of the tall, falcate dorsal fin is a good characteristic for identifying this species in the wild. (See Pl. 14.)

FIG. 26 Bottlenose Dolphin (*Tursiops truncatus*).

Distribution. Temperate and tropical waters. In all areas this species seems to be segregated into nearshore and offshore populations. The nearshore populations can be found in bays, harbors, lagoons, and estuaries, while the offshore populations range widely throughout the warmer oceans and seas of the world. Off California this dolphin is most common south of Point Conception, where it occurs nearshore and around all offshore islands. Migration patterns are not distinct, but seasonal north-south shifts in distribution regularly occur.

Natural History. This species most commonly travels in groups of 5 to 15 individuals, which, on occasion, join with other groups to form schools of several hundred animals. Each group tends to function as a single cohesive unit, demonstrating a great deal of intragroup helping and cooperative feeding behavior that suggests they may be composed of related individuals.

Captive Bottlenose Dolphins are often quite curious, and some individuals have demonstrated a remarkable ability to imitate the sounds and activities of other creatures. Free-ranging Bottlenose Dolphins show temerity towards human beings, and they have occasionally learned to exploit human activities for their own gain. For example, in certain areas Bottlenose Dolphins follow fish trawlers and shrimp boats, consuming some of the numerous fish and crustaceans disturbed by the dragging nets. This species readily rides ocean waves, from the pressure wave in front of vessels and large whales to breaking surf.

Bottlenose Dolphins are very vocal and undoubtedly communicate extensively among themselves. The sounds produced include ultrasonic echolocation clicks, high-frequency squeals and whistles, and low-frequency claps, grunts, and belches. The individual variation of the high-frequency whistle suggests that this sound functions as a personal signature.

Calving peaks in spring and summer following a gestation period of around 12 months. In captivity, and occasionally in the wild, one or more individuals have been observed acting very solicitously towards a calving mother and her neonate youngster. This helping behavior even extends to helping the newborn to the surface for its first breath. The calving interval is two to three years. At birth, calves measure somewhat over 1 m (3.5 ft) and weigh around 32 kg (70 lb).

Risso's Dolphin (*Grampus griseus*)

Description. Risso's Dolphin, commonly known as the Grampus, is a robust medium-sized delphinid that averages over 3.5 m (11 ft) in length and weighs about 300 kg (650 lb). Maximum reported size is nearly 4 m (13 ft) and 680 kg (1,500 lb). There is no distinct beak, although the jaw and upper lip project slightly forward. Anterior to the tall, falcate dorsal fin the body appears quite bulky for a delphinid. The posterior portion of the body displays the typical fusiform taper characteristic of most cetaceans. The long, pointed pectoral fins appear almost scythelike. The initial gray coloration of the body turns darker as the animal ages, but upon reaching full maturity the body begins to lighten again, turning a whitish cream color in older animals. All adult individuals show extensive scarring along the back and flanks. The Grampus rarely possesses visible teeth in the upper jaw, and the lower jaw typically contains only three to five pairs. (See Pl. 16.)

FIG. 27 Risso's Dolphin (*Grampus griseus*).

Distribution. Cosmopolitan in temperate and subtropical waters. Distributed from Alaska south to Central America in the North Pacific. The Grampus shies away from ocean-going vessels and is rarely sighted at sea. The lack of sightings led many marine biologists to conclude incorrectly that this species is primarily distributed far offshore. However, recent aerial survey work has established that these dolphins actually range widely over continental shelf waters (depth to 200 m, or 660 ft). They do not migrate extensively, although their seasonal distribution patterns suggest they may move farther offshore and north during winter. Off California they are most abundant north of Point Conception, and their greatest nearshore densities are recorded in winter.

Natural History. The Grampus commonly travels in schools of 15 to 40 individuals. On occasion, hundreds of these schools coalesce into the same area and literally checkerboard the surface of the ocean. The social nature of this species is also demonstrated by its frequent association with Pacific White-sided and Northern Right-whale Dolphins and with Short-finned Pilot Whales. The Grampus used to be considered somewhat uncommon, but aerial survey data demonstrate that it actually ranks among the top five most abundant cetaceans off California. Its food consists mainly of squid.

Since this species infrequently strands and is not commonly encountered at sea, little is known of its reproductive biology. The scant data available suggest that newborn calves are born in winter and measure about 1.5 m (5 ft). At present it seems that most reproductive activity occurs when these dolphins are aggregated in large groups.

Pantropical Spotted Dolphin (*Stenella attenuata*)

This species has not been recorded off the California coast. However, it has been included in this field guide because (1) it is one of the most common and abundant dolphins in the eastern North Pacific; (2) it is the major species involved in the controversial tuna-dolphin problem; and (3) it is common just south of California off central and lower Baja California. (See Pl. 17.)

Description. The maximum recorded size is 2.57 m (8.5 ft) and 119 kg (260 lb). Adult males, which are slightly larger than females, average 2.2 m (7 ft) and 80 kg (175 lb). In adults the dark gray-brown dorsal surface and upper flank region are

FIG. 28 Pantropical Spotted Dolphin (*Stenella attenuata*).

covered with light-colored spots, while the light gray lower flank and ventral regions are covered with dark spots. The size and number of spots increases continuously throughout life, eventually becoming mostly confluent in older animals. Black pigment surrounds each eye and extends forward in a thin line towards the forehead. A broad dark stripe runs from the flipper to the base of the predominantly dark jaws. The dorsal fin is located midbody and is moderately falcate. Typically, between 35 and 45 pairs of teeth line each jaw.

Distribution. Worldwide in tropical and warm subtropical waters. In the eastern North Pacific coastal populations are common from southern Baja California to near the equator. Offshore populations occur throughout the Pacific and Indian Oceans. Migration patterns are poorly known, but both coastal and offshore populations move extensively on a seasonal and annual basis.

Natural History. Very gregarious. Coastal herds typically number 50 or more, and offshore groups often exceed 1,000. Pantropical Spotted Dolphins are commonly found in association with Spinner Dolphins (*S. longirostris*), and they willingly ride the bow wave of vessels. An acrobatic and energetic swimmer, this species has been clocked at burst speeds of 40 km/hr (22 knots) in captivity.

In the tropical Pacific, schools of Yellowfin Tuna (*Thunnus alalunga*) often travel beneath groups of Spotted Dolphins. This characteristic behavior of the tuna has been widely ex-

FIG. 29 Spinner Dolphin (*Stenella longirostris*).

ploited by commercial fishermen. Fishermen locate a group of dolphins and rapidly surround the group with huge purse seine nets. After the nets are pursed (drawn closed), the dolphins, as well as the tuna below, are effectively trapped. During the process of retrieving the nets to collect the tuna, literally hundreds of thousands of dolphins have been inadvertently drowned. In recent years changes in fishing techniques, combined with quotas on the number of dolphins that can be killed by the tuna industry, have greatly reduced this tragic destruction by U.S.-based fishing boats. However, many foreign-based fleets still practice the old techniques and unnecessarily slaughter thousands of dolphins each year.

Calving is thought to occur throughout the year. Gestation lasts nearly one year, and lactation can last well over one year. The unspotted newborn calf usually measures somewhat less than 1 m (2.8 ft).

Striped Dolphin (*Stenella coeruleoalba*)

Description. Maximum length 2.5 m (8 ft), with males being slightly larger than females. Maximum weight 115 kg (250 lb). Jaws form a conspicuous beak in front of the gently sloping forehead. Dorsal fin and flippers are prominent and falcate. The dorsal surface and beak are dark bluish gray, while the chest and belly are white. A dark triangular-shaped region extends down and forward from behind the dorsal fin onto the light gray flanks. From the dark eyes three stripes extend posteriorly. A dark bold stripe runs directly to the flippers, a lighter stripe fades away just above the flippers, and a dark stripe terminates at the anus. Approximately 50 pairs of conical inwardly curved teeth line each jaw.

FIG. 30 Striped Dolphin (*Stenella coeruleoalba*).

Distribution. Offshore in temperate and tropical waters. Unusual off California. Stranding records suggest that this species occurs north to British Columbia during warm-water years. Abundant in tropical waters and in the western North Pacific, where it was formerly known as *S. styx*. Migration patterns are not well known, but it appears likely that the Striped Dolphin moves north and south seasonally.

Natural History. Gregarious, typically found in groups exceeding 100 dolphins, and occasionally schools numbering over 2,000. These dolphins will ride the bow wave of boats, and they commonly leap clear of the water. This species does not do well in captivity. The diet consists of squid, crustaceans, and mesopelagic fish.

Reproductive patterns in eastern North Pacific populations are not well studied. In the western North Pacific mating and calving peak in winter and spring. Gestation lasts about 12 months, and calves nurse well over one year. The calving interval is every three years, and newborn dolphins measure about 1 m (3 ft).

Northern Right-whale Dolphin (*Lissodelphis borealis*)

Description. Males of this sleek, slender species average over 2.8 m (9 ft) in length and weigh around 70 kg (150 lb). The very largest specimens may exceed 3 m (10 ft) and 90 kg (200 lb). Females are somewhat smaller. The lack of any ridge or dorsal fin on the smooth shiny black back uniquely

FIG. 31 Northern Right-whale Dolphin (*Lissodelphis borealis*).

identifies this species of dolphin in the North Pacific; it shares this character with the Right Whale. The beak is well defined and marked by a small white spot near the tip of the lower jaw. There are numerous small, conical teeth in both jaws. The upper surface of the flukes is medium gray, while the lower surface is white-gray. From below, a bright white hourglass-shaped region extends posteriorly from the chest to the flukes. The flippers are narrow, pointed, and curve sharply towards the rear of the animal.

Distribution. North Pacific; along the west coast of North America from at least British Columbia to San Diego. Although Northern Right-whale Dolphins generally occur in relatively cool waters, their population peaks off central and northern California in winter when water temperatures are near their annual high. Since this species has been infrequently identified at sea, it has long been assumed to be a rather uncommon deepwater form. However, aerial studies have revealed that it is actually quite common both inshore and offshore. Inshore population peaks seem to be correlated with availability of food and high degrees of reproductive activity.

Natural History. Like many delphinids, Northern Right-whale Dolphins are gregarious. Schools in excess of 3,000 animals have been recorded, and they are seldom sighted in schools numbering fewer than 20 individuals. These dolphins are observed as commonly in mixed species schools as they are by themselves. Their most frequent companions are Risso's and Pacific White-sided Dolphins, but they are known to occur with at least seven other species of marine mammals.

Northern Right-whale Dolphins are noted for being rapid swimmers, and they are capable of reaching burst speeds of 40 km/hr (22 knots). This species demonstrates a marked propensity to leap completely out of the water when traveling rapidly. At times the vast quantity of sea foam churned up by a large school of leaping dolphins is visible from several miles away.

Little is known concerning the food habits of these dolphins, but the available data suggest they are nighttime feeders on squid and lanternfish. For at least one population, calving and possibly mating occur in nearshore waters off central California in late winter. Newborns are creamy gray and measure about 1 m (3 ft).

Porpoises
Family Phocoenidae

The so-called true porpoises lack a beak, possess a low-slung, usually triangular dorsal fin or none at all, and their teeth are small and spade-shaped. Their bodies are short and stocky. There are three genera and six species in the family; two species are found off the California coast. They both frequent nearshore water, but Dall's Porpoise is also sighted well off-shore.

Harbor Porpoise (*Phocoena phocoena*)

Description. This is the smallest cetacean found in California waters, with adults generally measuring about 1.6 m (5 ft) and weighing less than 50 kg (110 lb). Maximum size is 2 m (6 ft) and 90 kg (200 lb). The compact body is slate gray to brownish above and light gray to white below. The dorsal fin is short and triangular and the flukes are slightly concave with a median notch. Each jaw contains 19 to 28 small, distinctly spade-shaped teeth.

FIG. 32 Harbor Porpoise (*Phocoena phocoena*).

Distribution. Relatively shallow nearshore waters throughout the ice-free zones of the North Pacific and North Atlantic. Generally preferring water cooler than 15° C (60° F), this species is somewhat less common south of San Francisco Bay and rarely strays into southern California waters. The name Harbor

Porpoise is somewhat of a misnomer, as it does not regularly occur in harbors. However, Harbor Porpoises often occur in bays and occasionally even in estuarine waters. They are not known to migrate to any great extent, and it seems likely that local resident populations occur all along the central and northern California coast.

Natural History. Harbor Porpoises off California are not well studied. These small, dark-colored cetaceans can be difficult to locate and nearly impossible to follow for extended periods. They typically expose only their dorsal fin and portions of their back when they surface to breathe, and they tend to occur in small groups in murky coastal waters. In addition, they actively avoid boats and rarely associate with other marine mammals or leap out of the water. Thus, they can be easily overlooked at the surface, and even if sighted they disappear upon submergence.

By frequenting shallow coastal waters, Harbor Porpoises become especially vulnerable to human activities. They have been, and in some areas still are, hunted for food, they often become entangled in fishing gear, and they are commonly exposed to highly toxic coastally discharged pollutants. Although their numbers appear to be declining, they are still considered fairly common from northern California to Alaska. The diet of this species consists of common nearshore invertebrates and schooling fish.

Mating occurs throughout the year with a possible peak in summer, as the majority of calves are born in late spring. Gestation lasts 10 to 11 months and weaning occurs at about 7 months. Newborn calves measure less than 1 m (3 ft) and weigh about 9 kg (20 lb).

Dall's Porpoise (*Phocoenoides dalli*)

Description. This is the largest member of the family Phocoenidae. The thicker-bodied, slightly larger male averages 1.8 m (6 ft) and 120 kg (270 lb) with a maximum size of 2.2 m (7 ft) and 220 kg (480 lb). The stout muscular body; small, triangular, predominantly white dorsal fin; and dramatic black and white body markings make this species distinctive. The jaws, which contain 20 to 30 small, slightly spade-shaped teeth, are

Plate 1. Gray Whale cow and calf.

Plate 2. Flukes of Gray Whale starting a dive.

Plate 3. Killer Whale. [Taken at Sea World, San Diego.]

Plate 4. Two Gray Whales, one blowing.

Plate 5. Aerial view of Gray Whale cow and calf.

Plate 6. Beached Gray Whale. [George E. Lindsay]

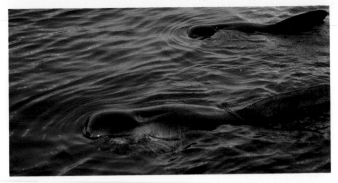

Plate 7. Profile of top of head and dorsal fin of two Short-finned Pilot Whales.

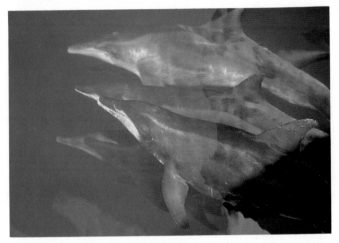

Plate 8. Underwater view of Rough-toothed Dolphins. [Marc Webber]

Plate 9. Common Dolphins leaping out of the sea.

Plate 10. Head of beached Hubbs' Beaked Whale. [Woody Williams]

Plate 11. Underwater view of a False Killer Whale (above) and two Short-finned Pilot Whales (below). [Courtesy of Marineland of the Pacific, Los Angeles.]

Plate 12. Common Dolphins in the Sea of Cortez. [George E. Lindsay]

Plate 13. Pacific White-sided Dolphins. [Courtesy of Steinhart Aquarium, San Francisco.]

Plate 14. Bottlenose Dolphin. [Courtesy of Steinhart Aquarium, San Francisco.]

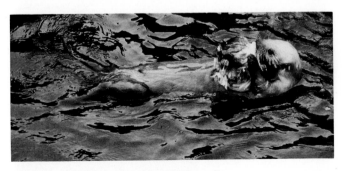

Plate 15. Sea Otter feeding. [Karl W. Kenyon]

Plate 16. Risso's Dolphin. [Marc Webber]

Plate 17. Pantropical Spotted Dolphin. [Marc Webber]

Plate 18. Dall's Porpoise
seen from above.
[Marc Webber]

Plate 19. Steller's Sea Lion bulls in foreground with female in the water.

Plate 20. Steller's Sea Lion bull surrounded by cows.

Plate 21. Female Steller's Sea Lions and pups.

Plate 22. Two aggressive Steller's Sea Lion bulls in the foreground with a dominant bull and cows behind. [Thomas C. Poulter]

Plate 23. Nursing Steller's Sea Lion pup and its mother.

Plate 24. Male California Sea Lions hauled out on Año Nuevo Island, California, during southern migration.

Plate 25. California Sea Lions hauled out on Año Nuevo Island during southern migration.

Plate 26. Bull Guadalupe Fur Seal on Guadalupe Island, Mexico. [George E. Lindsay]

Plate 27. Northern Fur Seal breeding colony, St. Paul Island, Alaska, showing difference in size between bulls and cows.

Plate 28. Male California Sea Lions hauled out.

Plate 29. Northern Fur
Seal bull, St. Paul Island,
Alaska.

Plate 30. Northern Fur Seal breeding rookery, St. Paul Island, Alaska.

Plate 31. Northern Elephant Seal "weaners" (newly weaned pups).

Plate 32. Immature Northern Elephant Seal during molt.

Plate 33. Harbor Seals hauled out on ledge close to water.

Plate 34. Northern Elephant Seal bulls fighting. [Roger C. Helm]

Plate 35. Breeding colony of Northern Elephant Seals on Año Nuevo Island, California. [Roger C. Helm]

Plate 36. Harbor Seals resting on rocky ledge.

not elongated into a prominent beak. The head is smallish, and the area just in front of the flukes is often peculiarly humped forming a caudal keel. When Dall's Porpoise swims rapidly only the upper portion of the body breaks the surface, resulting in a diagnostic "rooster-tail" splash. The form of the splash is such that an air pocket is formed across the forehead and back of the animal, thus allowing it to complete its breath before re-entry. The white-tipped flukes are a good field mark when the animals are viewed from the air. (See Pl. 18.)

FIG. 33 Dall's Porpoise (*Phocoenoides dalli*).

Distribution. Throughout the North Pacific; common from the Bering Sea to northern Baja California. Although Dall's Porpoises probably prefer colder waters, they are most common off California in autumn and winter. Along central and northern California they are common over continental shelf waters (depth to 200 m, or 660 ft), but as they move farther south they occur less commonly in nearshore waters. They do not seem to make long-distance migrations, but most subpopulations do show seasonal north-south and east-west shifts in distribution.

Natural History. Dall's Porpoises typically travel in small groups of 2 to 5 individuals, but occasionally schools in excess of 100 individuals are sighted. For most people, the first sighting of a Dall's Porpoise is thrilling. From several kilometers away a group of black and white forms are seen as they streak towards the bow of a boat. As they race along the surface, kicking up their characteristic rooster-tail spray, these porpoises can travel at speeds in excess of 35 km/hr (19 knots). Upon arriving at the vessel, the largest and presumably most dominant individuals jockey for favored positions within the pressure wave at the bow.

The wide variety of deepwater squid, fish, and crustaceans fed upon by Dall's Porpoise are known to rise towards surface waters each evening and return to deeper waters by daybreak. This fact, combined with the observation that Dall's Porpoises are typically sighted during the day slowly and rhythmically rising to the surface to breathe, as if sleeping, suggests that these porpoises feed nocturnally or in the early morning and late evening hours.

This species remains common off California despite heavy incidental kill of northern North Pacific populations by Japanese gill net fishermen and the harpooning of thousands of porpoises each year for human consumption by coastal Japanese whalers.

Dall's Porpoises are reproductively active year-round, with a summer peak in calving. The newborn young are approximately 1 m (3 ft) in length and weigh about 25 kg (55 lb). The gestation period is 10 to 12 months, and nursing may continue for up to two years.

3 · CARNIVORES
Order Carnivora

Carnivores are the major terrestrial group of flesh-eating mammals. However, not all carnivores eat flesh exclusively; raccoons and most bears are omnivores, eating both plant and animal food. One species, the Great Panda, is actually a herbivore whose main staple is bamboo. Although most carnivores are terrestrial, a few, like the Mink and river otters, are semi-aquatic; and in North America, one member of this order, the Sea Otter, has become strictly marine in habits.

Mustelids
Family Mustelidae

These are small to medium-sized carnivores that are characterized by well-developed scent glands. These glands may serve for individual and sex recognition, for marking territories, and for defense. The latter function is most highly developed in the skunks. The largest members of the family are the Wolverine and Sea Otter, and the smallest is the Least Weasel, whose body length is no more than that of a field mouse.

Sea Otter (*Enhydra lutris*)

Description. Length of adult males from 1.5 to 2 m (5 to 6 ft), females considerably smaller. Maximum weight recorded is 45 kg (100 lb). Pups average about 60 cm (2 ft) in length and weigh slightly over 2 kg (4.5 lb). Head relatively round compared with most other mustelids. Ears small and snout rather broad. Body long with tail broad at base but tapering and equal to about one-fourth to one-fifth of body length. Front feet comparatively small but hind feet large, webbed, and flipperlike. Fur extremely fine and dense, dark brown to nearly black when young but head and neck becoming frosted white with age, especially in males. Studies by Dr. Victor B. Scheffer have shown that an adult male Sea Otter may have 800 million fur fibers covering the body. This is said to equal twice the density of the fur on the Northern Fur Seal. (See Pl. 15.)

FIG. 34 Sea Otter (*Enhydra lutris*).

Distribution. Sea Otters originally ranged over an area nearly 9,600 km (6,000 mi) long, extending from the northern coast of Japan through the Kurile Islands to the Aleutian Islands of Alaska, then down the west coast of North America to islands along the west coast of central Baja California, Mexico. The fur trade in the eighteenth and nineteenth centuries nearly exterminated the species throughout its range. In recent years the northern population in Alaska has increased to well over 100,000 individuals because of protection, but the California population is still very small and restricted largely to the coastal area from San Luis Obispo County north to southern San Mateo County. Recently a number of Sea Otters have been translocated to San Nicolas Island off southern California, where it is hoped they will become established away from the main shipping route and the constant danger of an oil spill.

History. The history of the Sea Otter has been written many times because it is intimately connected with the exploration of western North America. The species was first discovered by the Russian explorer Captain Vitus Bering and his crew. They spent the winter on the Commander Islands in the Bering Sea, where their ship was forced to land while they were trying to return from the coast of America. Some of the men survived because the Sea Otter and other marine mammals provided them with food and clothing for protection against the rigorous subarctic climate. Bering himself died on the island, which now bears his name, but the surviving crew members returned to the China coast the following year. The fame of the Sea Otter soon spread and fabulous sums were being offered for its pelt.

Search for Sea Otter pelts for the European and Asian fur trade rapidly led to the exploration of the Aleutian Islands by the Russians. These Russian fur trappers gradually moved down the Pacific Coast to California, where they established Fort Ross in 1812. Sea Otters were reported to be especially abundant around the Farallon Islands, west of San Francisco Bay, and around the Santa Barbara Channel Islands off the southern California coast, and many thousands were taken in these areas during the first half of the nineteenth century. No fur-bearing mammal can withstand such a heavy harvest. This is especially true of a species whose distribution is limited to coastal areas and immediately adjacent offshore islands. The Sea Otter population first declined in Alaska, where hundreds of thousands of animals were killed. It is said that the almost complete elimination of Sea Otters off Alaska influenced the Russians to sell this land to the United States in 1867 for $7,200,000.

Before the beginning of the twentieth century, the Sea Otter was close to extinction and was completely eliminated over most of its formerly extensive range. In 1911 a treaty was signed by the United States, Russia, Japan, and Canada affording Sea Otters complete protection from commercial exploitation, and in 1913 the Aleutian Islands National Wildlife Refuge was established to provide additional safety for the few hundred animals remaining in Alaskan waters. Sea Otters by this time were gone from southeastern Alaska, British Columbia, and the coasts of Washington and Oregon. A small herd apparently survived along the rather inaccessible coast of Monterey County in central California and was "discovered" in 1938 when the new coastal highway connecting Monterey with San Simeon was opened. Through careful protection the Alaska population has made a strong comeback. In contrast, the California population has grown very slowly and presently numbers fewer than 2,000 animals. In recent years this population has shown very little growth. This has been attributed to several factors. Among these are drowning from entanglement in monofilament nets used close to shore and conflicts with the shellfish industry. Chemical pollution may also be another factor. Since the present limited range of the California Sea Otter population parallels the route used by oil tankers, this presents a very serious risk in the event of an oil spill. Because of these various problems the California population is currently on the Threatened Species List.

Natural History. The habits of Sea Otters are unique in many respects. Though they belong to a family of carnivores that are primarily terrestrial, they spend most of their lives in the sea. Sea Otters are seldom seen even on offshore rocks along the California coast. They float on their backs and also swim on their backs when moving on the surface of the water. Much of their time is spent in cleaning and grooming their dense, fine fur, which protects them from the cold ocean water. Unlike cetaceans and pinnipeds, they do not have an insulating layer of blubber beneath the skin.

When sleeping, Sea Otters often wind a strand of kelp about their bodies to prevent drifting away from the relative security of the inshore waters. A group of sleeping Sea Otters that have anchored themselves in this manner will all have their bodies oriented in the same direction because of currents or wind. When not sleeping or grooming, much of the rest of the time is devoted to feeding. These animals require a high intake of protein to maintain their body heat in a cold environment. Sea urchins, crabs, abalones, snails, and various other kinds of invertebrates are the principal food items consumed. Although some fishermen blame the small central California Sea Otter population for the reduction in the numbers of abalones, rather than overexploitation by people, records show that abalones were abundant along this coast in the early days when Sea Otters were present in vastly greater numbers than today.

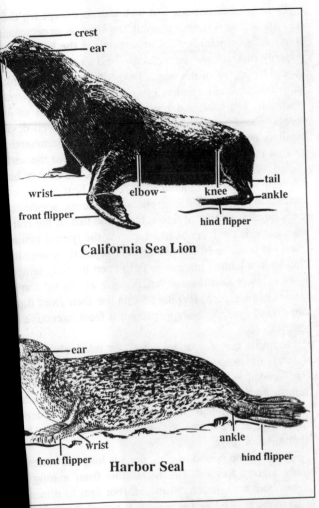

crest
ear

wrist
elbow
knee
tail
ankle
front flipper
hind flipper

California Sea Lion

ear

wrist
ankle
front flipper
hind flipper

Harbor Seal

Comparison of movement on land between an otariid seal (a Sea Lion) and a phocid seal (Harbor Seal); the California [Sea Lion] can move its hind flippers forward, but the Harbor Seal [cannot]

...t studies have shown that many kinds of pinnipeds are [capable] of producing underwater sounds. These may function [in comm]unication to some extent. It is possible, but not de[finitely pr]oved, that they may serve for echo ranging. If this is

Sea Otters, in a sense, make use of tools, as do only a very few other kinds of mammals apart from human beings. To assist in breaking up the hard exoskeleton of certain shellfish, they bring up rocks from the ocean floor and place them on their chest while floating on their backs. The prey is then pounded against the rock to crack the shell. Feeding occurs both during the day and at night.

There does not appear to be a definite reproductive season. Females attain reproductive maturity when between three and five years of age and may bear one young annually after a gestation period of six months. The pups are weaned when five to six months old, but they depend upon the female for some time after this until they have learned to capture their own food and use a tool. The males do not attain maturity until at least five years of age.

The principal enemies of Sea Otters, apart from human beings, are sharks, especially the Great White Shark, and Killer Whales, which are not often found in Sea Otter territory.

The best places to observe these fascinating animals are along the shore at Pacific Grove, Point Lobos State Park, along Seventeen-Mile Drive, and even from the Monterey wharf.

4 • SEALS, SEA LIONS, AND WALRUS
Order Pinnipedia

Seals, Sea Lions, and the Walrus belong to three distinct families. For many years they have been placed in an order called Pinnipedia on the basis of similarities resulting from adaptations to life in water. While there has never been any question as to the distinctness of the three families, some recent studies, based on fossil evidence, point to the greater resemblance each of these bear to certain families of carnivores than to each other. The eared seals and the Walrus show affinity with the bear family, while the earless seals may be allied with otters. This evidence, however, is contradicted by molecular studies, which seem to show that the genetic materials, the DNA, from the different families of pinnipeds are more similar to each other than they are to those of other carnivores. Hopefully, future studies of marine mammals will solve these problems of classification.

For practical purposes in this guide we are considering seals, sea lions, and the Walrus as members of a single order, the Pinnipedia. All are remarkably adapted to life in the sea. Their bodies are streamlined, the somewhat fusiform shape offering minimum resistance to passage through water. External ears are greatly reduced, with the pinna absent in the so-called earless seals. The eyes are large and well adapted to see under water. The limbs are modified into flippers by shortening of the bones of the front and hind legs, which are partly within the body, and elongation of the fingers and toes, which are joined together by connective tissue and skin to form broad propelling surfaces. The tail is extremely short.

Pinniped skin is thick and well-haired
rus, whose hair is sparse. Beneath the skin
layer of fat, which provides insulation fo
majority of seals and sea lions, as well a
the colder oceans of the world, the proble
temperature is an important one. Their in
essentially constant and about the san
beings. The skin and flippers are provide
capillary networks. When the body beco
passing rapidly through these tiny chann
cooled. When the body temperature dro
the peripheral areas is reduced and bod
temperature of the skin in arctic and a
times be close to that of the water, whil
animal inside the layer of blubber is at
level. On land the Northern Fur Seal r
flippers in a fanlike manner to help ke
down. This is also done when floatin
species of pinnipeds. Elephant Seals u
throw sand over the body to protect i
beaches.

Pinnipeds can stay under water fc
of time, and some species may mak
The Weddell Seal of the Antarctic ha
slightly more than 600 m (2,000 ft)
the Northern Elephant Seal has bee
again as great. This entails going w
30 minutes or more.

The gestation period in pinnipeds
erably among species. The total t
ranges from nine months in the Ha
in the Walrus. The actual growing
shorter because of delayed impla
from six weeks to five months. I
weeks in the Northern Elephant Se
sea lions. Rapid growth in young
content of the milk, which may be
fat content of the milk of otariids,
er, is lower.

FIG. 35
(Californi
Sea Lion

Recen
capable
for comn
finitely p

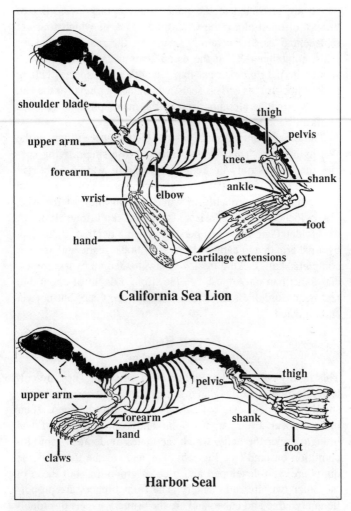

California Sea Lion

Harbor Seal

cannot and has to hop on its belly on land. The drawings on this page show the skeletal arrangement of the limbs in the two groups.

so, these underwater sounds could aid in orientation and location of food at night or at depths where visibility is limited, using a method similar to that used by cetaceans and bats.

There are over thirty living species of pinnipeds, most of

which are restricted to the polar and subpolar areas. Six of these are found along the California coast, of which four are representatives of the family Otariidae, the sea lions and fur seals, sometimes called the eared seals because they have a small external *pinna,* or ear flap. The other two species belong to the Phocidae, or earless seals, which lack a pinna for the ear. Fig. 35 provides a comparison of the major physical features of otariids and phocids.

The other pinniped family is the Odobenidae, to which the Walrus belongs. This is now restricted to a single living species that occurs only in arctic and subarctic waters. Subfossil evidence indicates that Walruses may have occurred along the California coast, probably in winter, during the last Ice Age. The evidence to date consists of a partial skull dragged up off the Golden Gate in November 1965 from a depth of 3,300 feet by a fisherman and carbon dated at 27,000+ years, and another complete skull, lacking the lower jaw, hauled up by a commercial fisherman on August 1, 1986, from a depth of about 340 feet near Cordell Bank, also off the Golden Gate, but not yet carbon dated.

Eared Seals
Family Otariidae

This family contains the sea lions and fur seals. They are collectively referred to as eared seals because all members have small but distinct external ears that can easily be seen. Their hind flippers are relatively long and can be reversed and brought under the body, which facilitates moderately rapid locomotion on land (see Fig. 35). The nails of the three middle digits are well-developed and used in grooming, but those of the outer two are rudimentary. The front flippers are proportionately large and naked and are the primary means of propulsion in the water. The small nails are not used in grooming. There is an extension of the flippers beyond the tips of the digits that is supported by cartilage.

Eared seals are found in the subpolar regions of both hemispheres as well as in temperate and even subtropical waters. They tend to be coastal in distribution, but the Northern Fur Seal is pelagic during part of the year. All species have a well-developed social system during the breeding period when they come to land. Only mature, dominant bulls acquire females. There is marked sexual dimorphism in size in all members of the family, the males in some species weighing up to five times as much as females. The food of eared seals consists primarily of available fish, ranging in size from herring to salmon. The four species that have been recorded along the California coast are Steller's or the Northern Sea Lion, the California Sea Lion, the Guadalupe Fur Seal, and the Northern Fur Seal.

Steller's Sea Lion (*Eumetopias jubata*)

Description. Average length of males 2.9 m (9.5 ft), females 2.4 m (7.5 ft), pups 1 m (3.3 ft). Mature males weigh 1 metric ton, females 270 kg (600 lb), pups 20 kg (44 lb). Color in both sexes yellowish, varying from almost cream color to yellowish brown. Neck of males very thick with hairs longer and coarser than on the rest of the body, forming a sort of mane. Top of head lacks the prominent crest seen in mature male California Sea Lions. Front and hind flippers blackish and naked. (See Pls. 19–23.)

Distribution. Widely distributed around both sides of the North Pacific, ranging from Hokkaido, Japan, north and east across the Bering Sea, then south along the eastern Pacific Coast to the Channel Islands of southern California. The northernmost breeding colony is on the Pribilof Islands, and the center of abundance is in the Aleutian Islands. The southernmost breeding colony is on Año Nuevo Island off southern San Mateo County, California. Up to 1977 this species regularly bred in small numbers on San Miguel Island of the Channel Island group.

History. At present, the world population of Steller's Sea Lions probably exceeds 250,000. The population count before the middle of the last century is thought to have been much higher, but intensive sealing for these animals began around 1860 and many colonies were completely eliminated. In central California skeletal remains indicate some breeding population on the Farallon Islands; prior to the turn of the century they even bred on Seal Rocks off San Francisco. It is only in

FIG. 36 Steller's Sea Lion (*Eumetopias jubata*), bull, cow, and pup.

recent years that breeding rookeries have become reestablished on the Farallons. Steller's Sea Lions, especially the large males, were killed for their hides, blubber, and reproductive organs, the latter considered by some to have aphrodisiac qualities. This killing of sea lions was not confined to California but occurred north to Alaska.

In recent decades the slaughter of sea lions for commercial purposes has ceased, and populations in some areas have increased. In other areas, noticeable declines, due possibly to disease and shooting, have occurred. A census taken in 1977 in the eastern Aleutian Islands of Alaska showed that the population had declined to half the number found there twenty years earlier. On Año Nuevo Island the population had declined from over 2,600 individuals, including young, during the peak of the breeding season in July 1963 to less than half this number in the summer of 1986.

Natural History. Unlike the California Sea Lion, the Steller's Sea Lion does not habitually enter bays, estuaries, or river mouths. It is a species of the outer coast, hauling out on rugged offshore rocks and rocky islands.

The social behavior of Steller's Sea Lion resembles in some respects that of the Northern Fur Seal, although there are many differences between the two species. On Año Nuevo Island the adult males begin arriving in early May, with maximum numbers present by late June or early July. They start leaving the island in late July and nearly all are gone by the beginning of September. The dominant bulls begin establishing territories on the seaward reefs almost as soon as they arrive. They engage in considerable aggressive behavior, and fierce battles take place. Only the most powerful and dominant individuals secure the most favorable positions. Defeated bulls remain on the periphery or in areas where there are no females. Subadult males or bachelors gather in areas away from the adult bulls. Here they may engage in a sort of mock battle with one another.

Adult bulls face each other when fighting and lunge at the head and neck of the opponent. Their powerful teeth and jaws can do great damage, and most dominant bulls show numerous scars of battle. Defeated bulls may have many open wounds on head and neck. When defending territory the bulls roar, with a sound quite different from the staccato bark of the California Sea Lion. This roar, combined with the comparatively long manelike hair on the neck, may account for the name "sea lion." Territorial males may remain in their territories without going to sea to feed for as long as two months. Females, however, regularly leave their young to feed.

Although some females, or cows, are present on Año Nuevo Island all year, a great influx occurs in late May and June, with peak numbers in July and August. Aggregations of females begin to form in late May around a single dominant male who may acquire as many as thirty cows, although the number is usually much smaller. Bulls try to keep cows from straying out of their own territories, which may be no more than 6 m (20 ft) in diameter, but unlike Northern Fur Seal bulls they do not try to herd them forcefully.

The females bear their single young any time from about June 1 to July 15. The fur of the pups is a dark grizzled brownish gray, which looks almost black when wet. Their calls remind one of the bleat of a lamb, and these vocalizations seem to be important in helping females locate their young. During the first week or two after birth there is fairly high mortality among the young. Some of the pups are crushed by the big bulls, who pay no attention to them, and some are swept off the reefs by rough seas. The young can swim but often are unable to get back up on the reefs, or else they swim away and become lost. When they are several weeks old they tend to sleep in groups on the edge of the breeding areas, where it is safer. Much of their waking time is spent swimming in tide pools. By the time they are about two and a half months old they spend much of the day in surge channels and swim between the rookery reefs. The young nurse for many months, and it is not uncommon to see females with yearlings and tiny pups both nursing.

California Sea Lion (*Zalophus californianus*)

Description. Average length of mature males 2.25 m (7.5 ft), females 1.8 m (6 ft), and pups 72 cm (2.3 ft). Average weight for males 325 kg (716 lb), females 110 kg (242 lb), pups 8 kg (18 lb). General color of males dark brown, almost black when wet. Females usually considerably lighter. Neck of male very thick. Top of head of mature males has a prominent crest covered with hair that is lighter than on the rest of the body. Both front and hind flippers are naked and blackish in color. (See Pls. 24, 25, and 28.)

Distribution. The breeding range of the California Sea Lion extends from the Channel Islands off southern California south along the coast of Mexico, possibly to the Tres Marias Islands. A number of islands in the Sea of Cortez have breeding populations, but the most important sites seem to be San Miguel

FIG. 37 California Sea Lion (*Zalophus californianus*), bull, cows, and pups.

and San Nicolas Islands in the Channel Island chain along with several islands off the Pacific Coast of Baja California. Occasionally pups are also born on Año Nuevo and the Farallon Islands.

Towards the end of the breeding season both adult and immature males start moving north from the west coast of Baja California and the Channel Island rookeries. This migration reaches its peak along the central and northern California coastline in September as counts of animals sighted at sea or hauled out reach their annual maximums. A count on Año Nuevo Island on August 30, 1963, showed over 13,000 California Sea Lions present. Nearly all of these were adult or immature males. Just 45 days earlier, on July 12, 1963, only 15 individuals were found on the island. Gradually many of these sea lions continue farther north to other hauling-out grounds along the Pacific Coast, where they are regular winter visitants up as far as British Columbia. There have even been rare sightings in Alaska.

The southward migration begins in early spring, and by late March or early April the population again increases along the coast of central California as the adult males, and some of the immatures, move south toward the breeding rookeries. By June most of the adult males are gone from northern and central California. However, immature California Sea Lions remain relatively abundant along the central California coastline during this time. The increasing numbers of these sea lions found off central and northern California throughout the year reflect the rather rapid population growth of these animals since the early 1970s. In contrast to the males, most of the Pacific Coast females stay around the rookeries all year. As far as is known, the Sea of Cortez population is nonmigratory, although there is some seasonal movement within the Gulf.

There are two other geographical forms of this species. One of these, *Zalophus californianus japonicus,* formerly inhabited some of the small islands off Japan and Korea but is now generally believed to be extinct. The other is *Z. c. wollebecki,* which occurs in the Galapagos Islands.

Natural History. The California Sea Lion is one of the best-known pinnipeds and is probably seen by more people than any other species in this group. These sea lions are common in zoos and are the so-called trained seals that one sees in circuses. They adapt well to captivity and are easily trained to perform.

Along the Pacific coast, California Sea Lions are seen on offshore rocks, secluded beaches, in bays, even on buoys. Some enter the lower parts of rivers, especially when there are fish migrating upstream. Their dark coloration, the high crest, and the loud staccato bark of the males readily distinguish them from Steller's Sea Lions, with which they are often associated on the outer coast.

In the water California Sea Lions are quite curious. They often come close to fishing boats and readily investigate the activities of skin and scuba divers. They are rapid and skillful swimmers, making use of their large front flippers for propulsion. At times, when trying to make speed, they "porpoise" on the surface and leap out of the water. They also seem to enjoy riding the waves as a surfer does. Individuals spend much time resting on sandy beaches or flat reefs of offshore islands, though they may also haul out on rather inaccessible beaches along the mainland, where they can be seen by observers but not easily bothered. Such areas are to be found along the Monterey coast and on Point Reyes in Marin County. Groups of individuals may rest on the surface of the water in rafts both during the day and at night. At such times a flipper is often raised into the air.

The breeding season for the California Sea Lion is June through July. The bulls become strongly territorial and aggressive at that time, and those that are dominant establish territories. These are adjacent to water and may be patrolled largely in the water by the bulls. Defense consists of nearly constant barking, threatening movements towards intruding bulls, and occasionally violent fighting. Territories are held for a bit less than one month on average.

Females are gregarious at all seasons of the year. They do not necessarily stay with the same bull for any length of time. A single male may have as many as fourteen females at a time in his territory. The females have a single young, usually born between late May and the middle of July. The female is bred about two weeks after the young is born. Females tend their young carefully for the first few days following birth, but the pups grow rapidly and within several weeks tend to gather in aggregations much like Steller's Sea Lion pups. The females continue to nurse them for many months and may do so for a year. During the early weeks of life the pups play in shallow water and on land, but as they grow older they venture farther out to sea.

During migration and in winter California Sea Lions are often closely associated on land with Steller's Sea Lions and Northern Elephant Seals. Although each species shows preference for a different habitat, there is much overlap because of limited shoreline space. California Sea Lion males mingle freely with female and young Steller's Sea Lions when they come together after the summer breeding season, and little if any aggression between the species is shown. They often sleep side by side on rocky ledges and reefs. Elephant Seals show little resistance to California Sea Lions sleeping on or climbing over their backs after the winter breeding period for the larger species is over. Harbor Seals show more caution but are also often found with these sea lions.

As with most pinnipeds, estimating the current population size of a species is difficult due to the relative inaccessibility of breeding rookeries and haul-out areas, the wide distribution of animals at sea, and the tendency of many members of the population to migrate long distances. However, data recently compiled during a multiyear aerial survey program suggest that over 150,000 California Sea Lions live along the west coast of North America and that somewhat less than half this total regularly occurs along the California coast. Furthermore, it appears that the sea lion population off California is growing at an annual rate in excess of 5 percent.

It seems unlikely that the California Sea Lion population will continue to increase at this rapid rate, considering the large number of factors that could increase the mortality rates within the population.

In certain areas Killer Whales and Great White Sharks may be responsible for some mortality. Commercial fishermen and their nets also pose serious problems for these animals locally, as does pollution. Lungworm, various intestinal parasites, and bacterial infections like *Leptospirosis* frequently debilitate these sea lions, increase the chance of reproductive failure, and often lead to death. In addition, the ever-increasing exploitation of fishery resources by both man and marine mammals will undoubtedly reduce the total amount of available prey.

Guadalupe Fur Seal
(*Arctocephalus townsendi*)

Description. Total length, males approximately 1.8 m (6 ft), females considerably smaller, 1.4 m (4.6 ft), and pups 60 cm (2 ft); weight, males about 160 kg (350 lb), females 45 kg (100 lb), pups 3 to 4 kg (7 to 9 lb). Muzzle proportionately long and pointed, with head rising rather abruptly. Front and hind flippers proportionately long. General color of thick fur a rich chestnut, with whitish tips to the coarser hairs producing a somewhat grayish or grizzled effect. (See Pl. 26.)

FIG. 38 Guadalupe Fur Seal (*Arctocephalus townsendi*), bull, cow, and pup.

Distribution. The present population of Guadalupe Fur Seals, numbering at least 2,000, is almost entirely confined to Isla de Guadalupe off the west coast of Baja California, although there are a few animals occasionally sighted on San Miguel and San Nicolas Islands.

History. This species was believed to have been common in the early days of sealing as far north as the Farallon Islands. Between 1810 and 1812, according to records, 73,402 skins of fur seals were taken at the Farallons. Although the sealers did not distinguish between the Northern Fur Seal (*Callorhinus ursinus*) and this more southern species, the animals were taken at a time of year when, according to later investigators, Northern Fur Seals from Alaska would have been absent or scarce. It was presumed, therefore, that most of those taken were Guadalupe Fur Seals. However, the fact that Northern Fur Seals have been successfully breeding on San Miguel Island since the 1960s casts serious doubt on this theory.

Natural History. There are eight southern species of fur seals of the genus *Arctocephalus.* Among the animals in this group, only the Guadalupe Fur Seal is distributed any distance north of the equator. The Galapagos Fur Seal breeds a few miles north of the equator on Isabella Island.

The little information available on the habits of this species suggests that at present it tends to spend much time in sea caves where it is difficult to observe. This habit may have been a factor responsible for the survival of a small population in spite of the inroads of the early fur sealers.

The bulls secure territories in or near sea caves and acquire a group of about ten females. The young are born in June, and July and after several weeks spend much time playing in tide pools. The common call of the males is described as a bark, but it differs from that of California Sea Lions or Northern Fur Seals. They may also utter a high-pitched roar when disturbed—even the young do this—but the sound is different from the calls of any other northern members of the family Otariidae.

The rediscovery in the early 1950s of this species, thought to be extinct, and the subsequent slow but constant increase in its numbers provides hope that the Guadalupe Fur Seal may come back as a significant member of California's marine mammal fauna.

Northern Fur Seal (*Callorhinus ursinus*)

Description. Total length, males slightly more than 2 m (7 ft), females 1.5 m (5 ft), pups 65 cm (2 ft); weight, males 180 to 270 kg (400 to 600 lb), females 43 to 50 kg (96 to 112 lb), pups 5 kg (10 to 12 lb). Nose pointed but muzzle short and profile of head decidedly convex from nose to neck. Adult male dark brown with neck and shoulders somewhat grizzled. Adult female dark gray above, lighter beneath. Hind flippers proportionately very long. (See Pls. 27, 29, and 30.)

Distribution. Breeds on the Pribilof and Commander Islands in the Bering Sea, the Robben and Kurile Islands near Japan, and San Miguel Island, California. In late summer, as the breeding season draws to an end, adult females from the Bering Sea population begin migrating south as far as Baja

FIG. 39 Northern Fur Seal (*Callorhinus ursinus*), bull, cows, and pup.

California. These animals remain well offshore and between winter and early spring are the most abundant pinniped in central and northern California.

History. Although few persons are aware that Northern Fur Seals are common off California in winter, this fact played an important part in the early exploration of the Pacific Coast. The first European scientist to observe this species was Georg Wilhelm Steller, who accompanied Vitus Bering in his search for Alaska. On August 10, 1741, he sighted what are believed to have been Northern Fur Seals. Steller again observed the species the following summer on Bering Island, where he was shipwrecked. In 1751 he published a description of this animal, and it was given its scientific name in 1758 by the great Swedish naturalist Linnaeus.

During succeeding years the Northern Fur Seal, sometimes called the Alaska or Pribilof Fur Seal, became well known to sealers. Millions of these animals were slaughtered for their fur, which is fine and thick, unlike that of sea lions. Sealing was carried out not only on the principal rookery islands, but on the open ocean, where the seals spend much of the year. It is said that the Russians took more than two and a half million Northern Fur Seal pelts on the Pribilof Islands between the time Pribilof discovered the breeding rookeries in 1786 and the purchase of Alaska by the United States from Russia in 1867. Subsequently, legislation set the islands aside as a reservation for the Northern Fur Seals, and regulations were established governing the taking of animals on these islands. Unrestricted sealing, however continued on the high seas and, since the majority of the fur seals taken by pelagic sealers were pregnant females or females with dependent pups on shore, and a great many of those shot were lost, the effect on the herd was serious. A multilateral convention was finally concluded in 1911 among Great Britain, Japan, Russia, and the United States prohibiting pelagic sealing, except by aborigines, and regulating the harvest on the rookery islands. The treaty was terminated by Japan in 1941, but in 1957 a new convention was signed, similar to that of 1911, by Canada, Japan, the U.S.S.R., and the United States. This treaty provides for the selective harvest of over 60,000 individuals annually of the estimated 1.8 million Northern Fur Seal population.

Natural History. The economic value of the Northern Fur Seal has led to intensive studies by biologists, and more is known about their habits and movements than about any other pinniped. Until the summer of 1968 the known breeding rookeries were the Pribilof Islands off Alaska, the Commander Islands off Kamchatka, Robben Island off Sakhalin, and some of the Kurile Islands. In June 1968, however, a group of about 100 individuals was observed on the shore of San Miguel Island off southern California. Forty of these were newborn pups, 60 were females, and one was a bull. Several breeding groups were found there in 1969 and again in 1970. Since then the San Miguel Island population has increased to around 10,000

Sea Otters, in a sense, make use of tools, as do only a very few other kinds of mammals apart from human beings. To assist in breaking up the hard exoskeleton of certain shellfish, they bring up rocks from the ocean floor and place them on their chest while floating on their backs. The prey is then pounded against the rock to crack the shell. Feeding occurs both during the day and at night.

There does not appear to be a definite reproductive season. Females attain reproductive maturity when between three and five years of age and may bear one young annually after a gestation period of six months. The pups are weaned when five to six months old, but they depend upon the female for some time after this until they have learned to capture their own food and use a tool. The males do not attain maturity until at least five years of age.

The principal enemies of Sea Otters, apart from human beings, are sharks, especially the Great White Shark, and Killer Whales, which are not often found in Sea Otter territory.

The best places to observe these fascinating animals are along the shore at Pacific Grove, Point Lobos State Park, along Seventeen-Mile Drive, and even from the Monterey wharf.

4 · SEALS, SEA LIONS, AND WALRUS
Order Pinnipedia

Seals, Sea Lions, and the Walrus belong to three distinct families. For many years they have been placed in an order called Pinnipedia on the basis of similarities resulting from adaptations to life in water. While there has never been any question as to the distinctness of the three families, some recent studies, based on fossil evidence, point to the greater resemblance each of these bear to certain families of carnivores than to each other. The eared seals and the Walrus show affinity with the bear family, while the earless seals may be allied with otters. This evidence, however, is contradicted by molecular studies, which seem to show that the genetic materials, the DNA, from the different families of pinnipeds are more similar to each other than they are to those of other carnivores. Hopefully, future studies of marine mammals will solve these problems of classification.

For practical purposes in this guide we are considering seals, sea lions, and the Walrus as members of a single order, the Pinnipedia. All are remarkably adapted to life in the sea. Their bodies are streamlined, the somewhat fusiform shape offering minimum resistance to passage through water. External ears are greatly reduced, with the pinna absent in the so-called earless seals. The eyes are large and well adapted to see under water. The limbs are modified into flippers by shortening of the bones of the front and hind legs, which are partly within the body, and elongation of the fingers and toes, which are joined together by connective tissue and skin to form broad propelling surfaces. The tail is extremely short.

Pinniped skin is thick and well-haired except for the Walrus, whose hair is sparse. Beneath the skin, as in cetaceans, is a layer of fat, which provides insulation for the body. Since the majority of seals and sea lions, as well as the Walrus, inhabit the colder oceans of the world, the problem of regulating body temperature is an important one. Their internal temperature is essentially constant and about the same as that of human beings. The skin and flippers are provided with well-developed capillary networks. When the body becomes overheated, blood passing rapidly through these tiny channels near the surface is cooled. When the body temperature drops, the flow of blood to the peripheral areas is reduced and body heat conserved. The temperature of the skin in arctic and antarctic species may at times be close to that of the water, while that of the rest of the animal inside the layer of blubber is at the typical mammalian level. On land the Northern Fur Seal regularly waves its large flippers in a fanlike manner to help keep its body temperature down. This is also done when floating at sea by some other species of pinnipeds. Elephant Seals use their front flippers to throw sand over the body to protect it from excessive heat on beaches.

Pinnipeds can stay under water for relatively long periods of time, and some species may make extremely deep dives. The Weddell Seal of the Antarctic has been recorded at depths slightly more than 600 m (2,000 ft) beneath the surface, and the Northern Elephant Seal has been recorded at depths half again as great. This entails going without breathing for about 30 minutes or more.

The gestation period in pinnipeds is long but varies considerably among species. The total time from mating to birth ranges from nine months in the Harbor Seal to fifteen months in the Walrus. The actual growing period, however, may be shorter because of delayed implantation. This delay varies from six weeks to five months. Lactation ranges from four weeks in the Northern Elephant Seal to a year or more in some sea lions. Rapid growth in young phocids results from the fat content of the milk, which may be as high as 55 percent. The fat content of the milk of otariids, whose growth is much slower, is lower.

FIG. 35 Comparison of movement on land between an otariid seal (California Sea Lion) and a phocid seal (Harbor Seal); the California Sea Lion can move its hind flippers forward, but the Harbor Seal

Recent studies have shown that many kinds of pinnipeds are capable of producing underwater sounds. These may function for communication to some extent. It is possible, but not definitely proved, that they may serve for echo ranging. If this is individuals. The island is nearly 5,500 km (3,400 mi) southeast of the previously known breeding range, and the presence of fur seals there may be an indication that the species, prior to it's decimation by man, had a range not unlike that of Steller's Sea Lion. Little is known yet of the seasonal movements of the San Miguel Island population, but it is thought these fur seals remain in California waters throughout the year. Bands have been found on five females captured on San Miguel Island; four of the females were banded by U.S. personnel on St. Paul Island, and one was banded by U.S.S.R. biologists on Bering Island of the Commander Group.

The Northern Fur Seal is gregarious, but unlike most pinnipeds it generally comes to land only during the reproductive season. The adult males winter principally in the Gulf of Alaska, while the immatures and females spread southward along the Pacific Coast of North America. They are most abundant 50 to 110 km (30 to 70 mi) offshore.

In late April the bulls begin to move northward. On the Pribilof Islands, where about 80 percent of the Northern Fur Seals breed, they increase in numbers until mid-June. The older and more dominant bulls are the first to arrive at the rookery and by aggressive behavior establish territories, which they hold for 40 to 50 days. The females do not appear until June. Bulls in the center of the rookery tend to acquire the largest groups of females, but they actually have little or no control over the size of groups. The end result is that some bulls have breeding access to large numbers of cows, while others have access to only a few. Most of the females have their young during the first three weeks of July and are bred five to seven days after giving birth. A female will remain with her pup for about a week, following which she will regularly go to sea to feed. The first excursions to sea last four to five days, with subsequent trips lasting about eight days. During her absence the young survives on the rich milk it has consumed. The bulls do not leave the rookery from the time they arrive until nearly all the females have left in late July or early August. The immature or subadult males are not allowed onto the breeding territories and gather apart in bachelor groups upon their arrival between June and early August. Nursing continues until late fall, when the females leave and the young are on their own. They leave the islands by November and go to sea. A few return the following autumn for a short time, but the majority do not come back until the second or third year of life.

Earless Seals
Family Phocidae

This family contains the "true seals," commonly referred to as earless seals. Other names by which they are known are hair seals and phocids.

Unlike sea lions and fur seals, earless seals have hind flippers that cannot turn forward (see Fig. 35). This greatly limits their locomotion on land, since they are limited to hopping on their belly. Both sides of their flippers are covered with hair, and nails are well developed on both fore and hind digits. The front flippers are relatively small, unlike those of members of the family Otariidae, and play an insignificant part in swimming, which is accomplished by lateral, undulating fishlike movements of the posterior part of the body and use of the expanded hind flippers. Except for the elephant seals there is only a slight difference in size between males and females.

Most members of this family occur in arctic and antarctic regions, but there are a few species that inhabit tropical and subtropical waters. Of the ten living genera and eighteen species, two occur along the coast of California; the Harbor Seal and the Northern Elephant Seal.

Harbor Seal (*Phoca vitulina*)

Description. Total length 1.5 to 1.8 m (5 to 6 ft), pups 80 cm (2.6 ft); weight 60 to 75 kg (130 to 165 lb). Males tend to be larger than females. Head more rounded and muzzle shorter than in most species of seals. Front flippers small and well haired. Hind flipper fairly large and capable of being expanded for swimming. Color extremely variable, ranging from pale silver-gray with black to dark brown spots to completely dark brown or blackish with spots hardly discernible. (See Pls. 33 and 36.)

Distribution. In the North Pacific from Alaska south along the coast of North America to islands off the west coast of Baja

FIG. 40 Harbor Seal (*Phoca vitulina*), mother and pup.

California and along the Asiatic coast to China. In the North Atlantic from Greenland south to North Carolina and from Iceland south to the shores of the Netherlands and occasionally France.

Natural History. Harbor Seals are fairly common along the coast of California but are never found in large numbers. Their population along the state coast has been estimated to be around 17,000. They usually occur in groups ranging from a few individuals to as many as two or three hundred, but they do not exhibit the sort of social behavior that is so characteristic of sea lions, fur seals, and elephant seals. Harbor Seals often come into bays and estuaries and may be seen resting on sandbars at low tide. Along the outer coast they also tend to haul out on reefs or small offshore rocks at low tide. On islands they are frequently seen lying on rocky ledges in the late afternoon. Members of this species are very wary and are able to detect the approach of a human being from a considerable distance. They seldom venture far from the water, in which they take refuge at the first intimation of danger. Their movements on land are clumsy, effected by hopping on their bellies.

In the water Harbor Seals often move with their round heads above the surface so they can observe. They are curious and may follow a person, staying 45 to 90 m (150 to 300 ft) offshore, as one walks along a beach or a cliff above the sea.

Members of this species are notably silent in contrast to many other pinnipeds. They occasionally slap the surface of

Seal numbered fewer than 100 individuals, all on Isla de Guadalupe, Baja California. With protection this small herd has increased, and the species has gradually moved north to islands along the northwest coast of Baja California, then to the Channel Islands, and on to central California. In the winter of 1960–61 the first pups were born on Año Nuevo Island off San Mateo County and that population has continued to grow, reaching over 2,500 animals by the mid-1980s. The population has expanded to such an extent that in the 1970s animals started breeding on the adjacent mainland beaches. This area is now a major attraction to visitors during the winter breeding season. In the early 1970s these seals also started breeding on the Farallon Islands, and in 1986 nineteen pups were produced even farther north along the Point Reyes coastline.

The Northern Elephant Seal is exceeded in size among all pinnipeds only by its counterpart in the Southern Hemisphere, the Southern Elephant Seal (*Mirounga leonina*). The two species are very similar in appearance and habits, but *M. leonina,* which is primarily associated with subantarctic islands, attains a greater length and weight.

The Northern Elephant Seal is a social pinniped, but its social structure differs from that of other pinnipeds such as fur seals and sea lions, whose dominant males defend specific territories during the breeding season. With elephant seals the location of the females is the determining factor. Most adult males start appearing at the rookery sites in early December and remain there during the entire breeding season without going to sea to feed. There is considerable fighting and aggressive behavior between these males, and a linear dominance hierarchy is established such that the alpha male dominates all other bulls on the beach, the beta male ranks second to the alpha male, and so on down the line. Aggressive behavior may just consist of threatening an opponent, but violent battles frequently take place. Under these circumstances two males face each other close up, often with chests contacting and heads held high. Each male will then slash down on his opponent with the canine teeth, trying to gouge a wound in a soft tissue area. Occasionally these battles result in serious wounds and substantial loss of blood. However, most blows are absorbed in the thick, tough skin shield covering the throat and chest area. Ultimately the strongest or most persistent bull succeeds in driving his opponent away, often back into the sea. The head, large proboscis, and neck of older bulls often show many scars as the result of such fights. Dominant males vigorously vocalize at potential opponents by emitting rhythmic snorts through the inflated nose, which usually hangs down just in front of the open mouth. The pharynx behind acts as a sort of resonance chamber so as to produce a loud sound that carries a long distance.

Females arrive at the rookeries by the end of December and gather together on sandy or gravelly beaches. They give birth within a week, and over the next day or so are bred by the dominant bull that controls the beach area they selected. Subordinate males often stay around the periphery of the females and try to sneak onto a territory and mate with a cow in estrus. Usually the dominant bull detects the intruder, as the loud protestations of the cow being waylaid arouse him rapidly. In some cases, the number of cows in an area is simply too many for a single bull, and lower-ranking males gain access to those females on the periphery of the territory. In small rookeries the linear dominance hierarchy is so well established that a single male may be responsible for all of the mating, and even in large rookeries only a small number of males mate the vast majority of females. It is likely that most males never attain a status high enough in the dominance hierarchy to secure matings, and of those few males that achieve very high dominance, most are able to maintain a high rank for only one or perhaps two breeding seasons. As a rule females tend to be irritated by the proximity of other females and will viciously attack a strange pup that wanders nearby.

Pups are born with a thick coat of black, curly hair. They nurse for about four weeks, by which time their weight has more than quadrupled to around 155 kg (350 lb). They are subsequently referred to as *weaners*. After her pup is weaned, the female leaves the rookery and goes to sea. The pups start to molt their natal coat about the time they are weaned, and by seven weeks of age they have acquired a silvery gray coat of hair that is short and straight. They stay around the rookery until April or May and then go to sea. By this time their bodies have elongated and they have become proficient at swimming and diving, and they also must be very hungry, since they have been without food for at least two months. These youngsters return to land in September and do not leave again until early December, when the big bulls begin arriving.

In late spring there is an influx of adult females as well as immature elephant seals to the rookery areas to molt. These animals leave in early summer, when the adult and subadult males come ashore to molt. Molting in elephant seals involves not only loss of the old coat and its replacement by new hair, but loss of the outer layer of the epidermis. As a result small sheets of outer skin and hair may come off together. After the molt is over the males leave and are replaced by another influx of immatures, including young-of-the-year.

The whereabouts of Northern Elephant Seals when they go to sea is poorly known. Males tend to move north, and stray individuals have been seen as far north as Prince of Wales Island in Alaska. The limited information available suggests these seals stay fairly far from shore and tend to be solitary.

The food that has been found in stomachs of the few dead individuals that have washed ashore consists of squid, sharks, and fishes that live at considerable depths and are fairly slow swimmers. Recent research has demonstrated that some elephant seals may regularly dive to depths in excess of 700 m (2,300 ft).

The greatest mortality among pups results from being crushed by adults, especially bulls, or being bitten in the head by females. The latter are very aggressive when approached by a pup that is not their own. Natural mortality among first-year seals, as with most long-lived species, is high. Scars and wounds on the bodies of adults indicate that attacks by sharks, particularly the Great White Shark, regularly occur, and Killer Whales have been observed capturing and killing an adult bull.

Glossary

Amphipod: One of an order of small crustaceans living in or next to water, including sand fleas and whale lice; some species constitute a large part of the diet of baleen whales.

Baleen: Long plates of a horny material that hangs in rows from the roof of the mouth of members of the suborder Mysticeti. It functions as a food strainer.

Beluga: A name used for the sturgeon of the Black and Caspian Seas whose roe is used as caviar. Also applied to the White Whale of the Arctic.

Blow: In cetaceans this refers to the act of forcibly exhaling through the nasal opening or openings.

Blowhole: A nostril opening on the top of the head in cetaceans.

Blubber: A layer of fat and connective tissue beneath the skin of many marine mammals providing insulation from cold.

Breaching: Refers to whales leaping more or less vertically out of the water, thus exposing a large part of the body.

Bull: A term applied to adult male cetaceans and pinnipeds.

Calf: Refers to the young of some marine mammals.

Calving: Giving birth to young, as in cetaceans.

Carnivore: An organism that feeds primarily on animals. Also, a member of the order Carnivora.

Cephalopod: A member of Cephalopoda, the class of mollusks that includes squids and octopi.

Continental shelf waters: Waters adjacent to a continent and less than 200 m (660 ft) deep.

Continental slope waters: Waters that are approximately 200 to 2,000 m (660 to 6,600 ft) deep and lie between the point where the continental shelf ends and the deep ocean floor begins.

Cosmopolitan: Occurring widely over the world.

Cow: The adult female of some marine mammals.

Delphinid: Cetaceans of the family Delphinidae.

Diatoms: Unicellular algae possessing a hard cell wall.

Digits: Fingers and toes.

Distribution: Geographic range.

Drag: Resistance of the body to water.

87

Falcate: Hooked or curved like a sickle, usually in reference to the dorsal fin of a cetacean.

Flukes: Laterally flattened lobes formed by dorsoventral compression of the tail.

Gestation period: Period of pregnancy.

Gregarious: Tending to flock or herd together.

Grooming: Used here principally in reference to Sea Otters and fur seals cleaning their fur so that it functions properly for insulation.

Haul out: To move out of the water to a resting site on land.

Herbivore: Organism that primarily consumes plant material.

Invertebrate: Animal that lacks a backbone.

Knot: Unit of speed measurement at sea. 1 knot = 1 nautical mile per hour or approximately 1.15 statute miles per hour.

Krill: Planktonic pelagic crustaceans that constitute, along with larval and juvenile fish, the primary food of most baleen whales.

Lactation: The formation and secretion of milk by female mammals.

Mass stranding: The voluntary beaching of a large group of whales or dolphins.

Melon: Mass of adipose tissue in the snout region above the upper jaw of odontocetes.

Mesopelagic fish: Fish that live in midocean depths and typically rise towards surface waters each evening to feed.

Migration: A periodic movement between two regions by the majority of a population. The movement usually occurs along well-defined routes and typically culminates in feeding or breeding areas.

Morphologic: Refers to the form and structure of a living organism.

Mysticeti: Suborder of Cetacea comprised of whales with a symmetrical skull, paired blowhole, baleen, and lacking teeth. See **Odontoceti** and **Baleen.**

Nautical mile: International distance unit measurement for sea or air travel. Equals approximately 1.85 km or 1.15 statute miles. See **Knot.**

Neonate: Newly born, or newly born young.

Odontoceti: Suborder of Cetacea comprised of whales and dolphins with an asymmetrical skull, single blowhole, and teeth. See **Mysticeti.**

Omnivore: Organism that consumes both plant material and animal flesh.

Pantropical: Distributed throughout tropical areas.

Pectoral fin: Either of the paired forefins of an aquatic animal.

Pelagic: Inhabiting the upper waters of the open ocean.

Pinna: Projecting cartilaginous flap that deflects sound into the middle and inner ear of most mammals.

Pinniped: The mammalian group that includes the seals, sea lions, and Walrus.

Plankton: Freely drifting or weakly motile aquatic animals and plants that are the base of the marine food chain. Phytoplankton consists of plants, mostly diatoms. Zooplankton includes larval fish, but is mostly composed of minute crustaceans.

Pod: A group of cetaceans clustered together.

Porpoise: Cetacean of the family Phocoenidae characterized by a blunt snout and spade-shaped teeth. The ancient Greeks distinguished porpoises from dolphins (family Delphinidae). Vernacular use of the word "Porpoise" for "Dolphin" should be avoided as it leads to confusion. See text for further description of each group.

Rookery: Among pinnipeds the term refers to a location regularly used for breeding and rearing young.

Rorqual: Designates the whales of the family Balaenopteridae.

School: A large number of animals feeding, resting, or traveling together.

Sebaceous gland: A cutaneous gland in mammals that secretes an oily substance that prevents desiccation and enhances water-repellence of the skin and hair.

Sexual dimorphism: The notable differences in size, body shape, and color that exist in many species between mature males and females.

Sound: (1) To dive deeply, used especially when referring to large whales. (2) Ultrasonic sound: very high-frequency echolocation pulse produced by most odontocetes. (3) Subsonic sounds: any of a variety of sounds typically audible to humans.

Spermaceti: A fatty substance found in the heads of Sperm Whales.

Spyhopping: A characteristic behavior of certain whales (such as the Gray Whale) in which the animal rises vertically out of the water, exposing its eyes, and appears to search its surroundings visually.

Wean: Adult-enforced termination of the suckling period.

Weaner: A weaned pinniped pup.

Whale: Any of the Mysticeti or one of the larger Odontoceti, such as the Killer Whale, which is actually a large dolphin.

Whalebone: See **Baleen.**

Selected References

Baker, S., and L. M. Herman. 1984. Season contrasts in the social behavior of the Humpback Whale. *Cetus* 5(2):14–16.

Dairs, B. S. 1977. The Southern Sea Otter revisited. *Pacific Discovery* 30(2):1–13.

Earle, S. 1979. Humpbacks: The gentle whales. *National Geographic Magazine* 155(1):2–17.

Gentry, R. L. 1987. Seals and their kin. *National Geographic Magazine* 171(4):474–501.

Graves, W. 1976. The imperiled giants. *National Geographic Magazine* 150(6): 723–751.

Haley, D. (ed). 1986. *Marine Mammals of Eastern North Pacific and Arctic Waters* (2d ed. rev.). Seattle, Wash.: Pacific Search Press.

Horwood, J. 1987. *The Sei Whale: Population Biology, Ecology, and Management*. London: Croom Helm.

Hoyt, E. 1984. *The Whale Watcher's Handbook*. Illustrated by Pieter Folkens. A Madison Press Book. Garden City, N.Y.: Doubleday.

Jones, M. L., S. L. Swartz, and S. Leatherwood. 1984. *The Gray Whale: Eschrichtius robustus*. Orlando, Fl.: Academic Press.

Kaza, S. 1987. *Return of the Elephant Seal*. Point Reyes Bird Observatory, Spring 1987.

Kenyon, K. W. 1969. *The Sea Otter in the Eastern Pacific Ocean*. U.S. Dept. of the Interior. *North American Fauna 68 N*.

King, J. E. 1983. *Seals of the World* (2d ed.). British Museum (Natural History). Ithaca, N. Y.: Comstock Publishing Associates.

Leatherwood S., R. R. Reeves, W. F. Perrin, and W. E. Evans. 1982. *Whales, Dolphins, and Porpoises of the Eastern North Pacific and Adjacent Arctic Waters. A Guide to Their Identification*. NOAA Technical Report NMFS Circular 444.

Leatherwood, S., and R. R. Reeves. 1983. *The Sierra Club Handbook of Whales and Dolphins*. Illustrated by L. Foster. San Francisco: Sierra Club Books.

LeBoeuf, B. J., and S. Kaza (eds.). 1981. *The Natural History of Año Nuevo*. Pacific Grove, Calif.: The Boxwood Press.

Minasian, S. M., K. C. Balcomb, III, and L. Foster. 1984. *The World's Whales: The Complete Illustrated Guide*. Washington, D. C.: Smithsonian Books.

Payne, R. 1976. At home with Right Whales. *National Geographic Magazine* 149(3):322–339.

Perrin, W. J., R. L. Brownell, Jr., and D. P. DeMaster (eds.). 1984. *Reproduction in Whales, Dolphins, and Porpoises*. Proceedings of conference on cetacean reproduction, La Jolla, Calif., 1981. Reports of the International Whaling Commission, Special Issue 6.

Peterson, R. S., and G. A. Bartholomew. 1967. *The Natural History and Behavior of the California Sea Lion.* The American Society of Mammalogists Special Publication No. 1.

Radford, K. W., R. T. Orr, and C. L. Hubbs. 1965. Reestablishment of the Northern Elephant Seal (*Mirounga angustirostris*) off central California. *Proc. Cal. Acad. Sciences* 31(22):601–612.

Rice, D. W., A. A. Wolman, B. R. Mate, and J. T. Harvey. 1986. A mass stranding of Sperm Whales in Oregon: sex and age composition of the school. *Marine Mammal Science* 2(1):64–69.

Ridgway, S. H., and R. J. Harrison (eds.). 1981. *Handbook of Marine Mammals, Vol. 1: The Walrus, Sea Lions, Fur Seals, and Sea Otter.* New York: Academic Press.

Ridgway, S. H., and R. J. Harrison (eds.). 1981. *Handbook of Marine Mammals, Vol. 2: Seals.* New York: Academic Press.

Ridgway, S. H., and R. J. Harrison (eds.). 1985. *Handbook of Marine Mammals, Vol. 3: The Sirenians and Baleen Whales.* New York: Academic Press.

Scheffer, V. B. 1976. Exploring the lives of whales. *National Geographic Magazine* 150(6):752–767.

Shane, S. H., S. W. Randall, and B. Wursig. 1986. Ecology, behavior and social organization of the Bottlenose Dolphin: A review. *Marine Mammal Science* 2(1):34–36.

Sumich, J. L. 1984. Gray Whales along the Oregon Coast, 1977–1980. *Murrelet* 65:33–40.

Swartz, L., and M. L. Jones. 1987. Gray Whales: At play in Baja's San Ignacio Lagoon. *National Geographic Magazine* 171(6): 754–771.

Watson, L. 1985. *Whales of the World: A Complete Guide to the World's Living Whales, Dolphins, and Porpoises.* London: Hutchinson.

Woolfenden, J. 1985. *The California Sea Otter: Saved or doomed?,* rev. ed. Pacific Grove, Calif.: The Boxwood Press.

Index

Editor: Sean Cotter
Designer: Nancy Warner
Compositor: Warner-Cotter Co.
Text: 10/12 Times Roman
Display: Helvetica
Printer: Arcata Graphics/Fairfield
Binder: Arcata Graphics/Fairfield